Staff Responses to Frequently Asked Questions Concerning Decommissioning of Nuclear Power Plants

Final Report

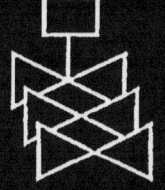

U.S. Nuclear Regulatory Commission
Office of Nuclear Reactor Regulation
Washington, DC 20555-0001

AVAILABILITY OF REFERENCE MATERIALS
IN NRC PUBLICATIONS

NRC Reference Material

As of November 1999, you may electronically access NUREG-series publications and other NRC records at NRC's Public Electronic Reading Room at www.nrc.gov/NRC/ADAMS/index.html. Publicly released records include, to name a few, NUREG-series publications; *Federal Register* notices; applicant, licensee, and vendor documents and correspondence; NRC correspondence and internal memoranda; bulletins and information notices; inspection and investigative reports; licensee event reports; and Commission papers and their attachments.

NRC publications in the NUREG series, NRC regulations, and *Title 10, Energy,* in the Code of *Federal Regulations* may also be purchased from one of these two sources.
1. The Superintendent of Documents
   U.S. Government Printing Office
   P. O. Box 37082
   Washington, DC 20402–9328
   www.access.gpo.gov/su_docs
   202–512–1800
2. The National Technical Information Service
   Springfield, VA 22161–0002
   www.ntis.gov
   1–800–553–6847 or, locally, 703–605–6000

A single copy of each NRC draft report for comment is available free, to the extent of supply, upon written request as follows:
Address: Office of the Chief Information Officer,
            Reproduction and Distribution
              Services Section
           U.S. Nuclear Regulatory Commission
           Washington, DC 20555-0001
E-mail: DISTRIBUTION@nrc.gov
Facsimile: 301–415–2289

Some publications in the NUREG series that are posted at NRC's Web site address www.nrc.gov/NRC/NUREGS/indexnum.html are updated regularly and may differ from the last printed version.

Non-NRC Reference Material

Documents available from public and special technical libraries include all open literature items, such as books, journal articles, and transactions, *Federal Register* notices, Federal and State legislation, and congressional reports. Such documents as theses, dissertations, foreign reports and translations, and non-NRC conference proceedings may be purchased from their sponsoring organization.

Copies of industry codes and standards used in a substantive manner in the NRC regulatory process are maintained at—
    The NRC Technical Library
    Two White Flint North
    11545 Rockville Pike
    Rockville, MD 20852–2738

These standards are available in the library for reference use by the public. Codes and standards are usually copyrighted and may be purchased from the originating organization or, if they are American National Standards, from—
    American National Standards Institute
    11 West 42nd Street
    New York, NY 10036–8002
    www.ansi.org
    212–642–4900

The NUREG series comprises (1) technical and administrative reports and books prepared by the staff (NUREG–XXXX) or agency contractors (NUREG/CR–XXXX), (2) proceedings of conferences (NUREG/CP–XXXX), (3) reports resulting from international agreements (NUREG/IA–XXXX), (4) brochures (NUREG/BR–XXXX), and (5) compilations of legal decisions and orders of the Commission and Atomic and Safety Licensing Boards and of Directors' decisions under Section 2.206 of NRC's regulations (NUREG–0750).

NUREG-1628

Staff Responses to Frequently Asked Questions Concerning Decommissioning of Nuclear Power Plants

Final Report

Manuscript Completed: May 2000
Date Published: June 2000

Prepared by
J.L. Minns, M.T. Masnik, NRC
R. Harty, E.E. Hickey, PNNL

Pacific Northwest National Laboratory
Richland, WA 99352

Division of Licensing Project Management
Office of Nuclear Reactor Regulation
U.S. Nuclear Regulatory Commission
Washington, DC 20555-0001

ABSTRACT

This report, through a question-and-answer format, provides U.S. Nuclear Regulatory Commission (NRC) staff responses to frequently asked questions on the decommissioning process for commercial, nuclear power reactors. The questions were taken from a variety of sources over the past several years, including written inquiries to the NRC and questions asked at public meetings and during informal discussions with the NRC staff. In responding to the questions, the NRC staff attempted to provide the answers in a clear and non-technical form.

With the increase in the number of power reactors beginning the decommissioning process and significant changes that occurred in the regulations since 1996, the staff realized that there was a general lack of understanding of the decommissioning process and the risks associated with decommissioning. This document was developed in response to the staff's concerns. The report contains a definition of decommissioning and a discussion of alternatives. It also provides a focus on decommissioning experiences in the United States and how the NRC regulates the decommissioning process. Questions related to spent fuel, low-level waste, and transportation related to decommissioning are answered. Questions related to license termination, the ultimate disposition of the facility, and finances for completing decommissioning and hazards associated with decommissioning are also addressed. This document also provides responses to questions related to public involvement in decommissioning as well as providing the public with sources for obtaining additional information on decommissioning.

Contents

ABBREVIATIONS

ADAMS	Agency-Wide Documents Access and Management System
ALARA	As low as is reasonably achievable
BWR	Boiling Water Reactor
CEDE	Committed Effective Dose Equivalent
CFR	Code of Federal Regulations
DOE	U.S. Department of Energy
DOT	U.S. Department of Transportation
EA	Environmental Assessment
EIS	Environmental Impact Statement
EPA	U.S. Environmental Protection Agency
EST	Eastern Standard Time
FRERP	Federal Radiological Emergency Response Plan
GEIS	Generic Environmental Impact Statement
GE-VBWR	General Electric - Vallicetos Boiling Water Reactor
GPO	Government Printing Office
HEPA	High-efficiency particulate air filter
HLW	High level waste
ICRP	International Commission on Radiation Protection
ISFSI	Independent Spent Fuel Storage Installation
LLNL	Lawrence Livermore National Laboratory
LLW	Low-level waste
NCRP	National Council on Radiation Protection and Measurements

NPDES	National Pollutant Discharge Elimination System
NRC	U.S. Nuclear Regulatory Commission
ODCM	Offsite Dose Calculation Manual
PSDAR	Post-shutdown Decommissioning Activities Report
PWR	Pressurized Water Reactor
SFP	Spent Fuel Pool
TEDE	Total Effective Dose Equivalent
TDD	Telecommunications device for the deaf
URL	Universal Resource Locator
U.S.	United States

ACKNOWLEDGMENTS

This report, which represents the cumulative efforts of many, was facilitated by bringing together a number of subject-matter specialists from within NRC to make comments on the document. We greatly appreciate the guidance, assistance, technical review, and encouragement provided by Mr. Tim Johnson, Ms. Sherry Lewis, Ms. Cynthia Sochor, Mr. Chris Gratton, Mr. Robert Wood, Mr. Brian Richter, Mr. Stephen Klementowicz, Ms Eileen M. McKenna, Ms. Elizabeth A. Hayden, Dr. M. Nalluswami, Mr. Barry Zalcman, Ms. Diane Jackson, Mr. James H. Wilson, Mr. George Mencinsky, Mr. Spiros Droggitis, Mr. Robert Lewis, Mr. Vincent J Everett, Mr. Mark Delligatti, Ms. Nancy Osgood, Mr. Patrick M. Madden, Mr. George Hubbard, and other NRC staff members. We also acknowledge the project managers of the Decommissioning Section: Mr. Thomas Frederichs, Mr. Duke Wheeler, Mr. Phillip Ray, Mr. Mike Webb, Mr. Dino Scaletti, Mr. John Hickman, Mr. Paul Harris, Mr. Dick Dudley, Mr. Lee Thonus, Mr. Mort Fairtile, Ms Ann P.Hodgdon, Mr. David J. Wrona, Mr. William Huffman and Mr. Stuart A. Richards, Director of Project Directorate IV & Decommissioning., A special thanks to Ms. Ray Sanders, Ms Barbara A. Calure, and Ms Elaine B Morris, Technical Editors.

INTRODUCTION

On July 20, 1995, the Commission issued a "Notice of Proposed Rulemaking on Decommissioning of Nuclear Power Plants." On July 2, 1996, the Commission approved the final rule. The rule was published in the *Federal Register* on July 29, 1996, and became effective 30 days from the date of publication (August 28, 1996) (61 FR 39278). The final rule (1) redefines the decommissioning process, (2) defines terminology related to decommissioning, (3) requires licensees to provide the NRC with early notification of planned decommissioning activities at their facilities, and (4) explicitly states the applicability of certain NRC requirements that are specific for reactors that are permanently shut down. On July 21, 1997, the NRC published in the *Federal Register* a final rule entitled, "Radiological Criteria for License Termination" (62 FR 39058). This rule contains the regulations regarding the radiological criteria that the licensee must meet before the license can be terminated. These regulations superseded regulations written in 1998.

The staff realized that there was a general lack of understanding of the decommissioning process and the risks associated with decommissioning. With the anticipated increase in the number of power reactors beginning the decommissioning process and the significant changes that occurred in the regulations since 1996, this report, through a question and answer format, provides NRC staff responses to frequently asked questions on decommissioning of commercial nuclear power reactors.

The questions were taken from a variety of sources over the past several years, including written inquiries to the NRC and questions asked at public meetings and during informal discussions with the NRC staff. In responding to the questions, the NRC staff attempted to provide the answers in a clear and non-technical form.

Sections 1 and 2 define decommissioning and discuss alternatives. Section 3 focuses on decommissioning experiences within the United States. Section 4 describes how the NRC regulates the decommissioning process. Sections 5, 6, and 7 concern spent fuel, low-level waste during decommissioning, and transportation, respectively. Sections 8 and 9, respectively, consider questions and answers on license termination, the ultimate disposition of the facility, and hazards associated with decommissioning. Section 10 addresses the financial aspects of funding decommissioning. Socioeconomic issues are discussed in Section 11. Section 12 discusses public involvement in the decommissioning process, with an emphasis on the early phases of decommissioning. Section 13 provides the public with sources of additional information on decommissioning. The final section contains a bibliography with relevant published materials.

1 GENERAL

1.1 How is decommissioning defined?

Title 10 of the *Code of Federal Regulations*, Section 50.2 (10 CFR 50.2) defines decommissioning as the safe removal of a facility from service and reduction of residual radioactivity to a level that permits termination of the NRC license.

1.2 Why do nuclear power plants shut down permanently?

Nuclear power plants cease operations for a variety of reasons. The NRC grants a license for a period of 40 years. At the end of the license period, the licensee can seek to renew the operating license of the plant for another 20 years, or can cease operations and begin the decommissioning process. Some licensees choose to cease power operations before the 40-year licensing period has been completed. Reasons for this decision are usually financial. For example, the plant may require upgrades or repairs that are not economically justifiable, or the licensee may find other sources of power that are less expensive than nuclear generation. In addition to financial reasons for decommissioning, the NRC can order the licensee to cease operations for safety reasons.

1.3 Why are power reactors decommissioned?

As one of the conditions for an operating license, the NRC requires the licensee to commit to decommissioning the nuclear plant after it ceases power operations. This requirement is based on the need to reduce the amount of radioactive material at the site in order to ensure public health and safety as well as the protection of the environment.

1.4 How does decommissioning proceed?

The regulations are written so that when a licensee announces its decision to permanently cease power operations at the nuclear power plant (or in extreme cases, when the NRC requires a licensee to cease operations), the decommissioning process is automatically initiated, and specific decisions regarding the decommissioning process must be made within 2 years. However, no major decommissioning activities can take place until the licensee has provided the NRC with specific information regarding the decommissioning process as required by the decommissioning regulations discussed later.

It is possible for the licensee to let the facility sit idle for a number of years before announcing its decision to permanently cease power operations (although the time could not extend beyond the duration of the operating license). However, it is not in the licensee's financial interest to delay this decision since the costs required to meet the regulations at an operating plant are much greater than the costs for a decommissioning plant.

1.5 What are the benefits of decommissioning?

The major benefit of decommissioning for the licensee as well as the public is that the levels of radioactive material at the site are reduced to levels that permit termination of the license and use of the site for other activities, rather than leaving the radioactively contaminated material on the site so that it could adversely affect public health and safety and the environment in the future.

1.6 What are the costs of decommissioning?

The major costs of decommissioning are the large financial costs involved in funding the project. Substantial costs are incurred in the removal, shipment, and disposal of major components of the facility that are contaminated, such as pumps, valves, piping, the steam generators, the pressurizer and the reactor vessel. Decontamination of floors, walls and equipment also result in substantial costs. The occupational dose received by workers during decommissioning should also be considered a cost.

1.7 What are the options to decommissioning?

NRC's regulations do not allow the option of not decommissioning. Under NRC regulations, the original operating license for a nuclear power plant is issued for up to 40 years. The license may be renewed for up to an additional 20 years if NRC requirements are met. However, at the end of the licensing period, the regulations require that the facility be decommissioned. The alternative to decommissioning, at the end of the licensing period, is a "no action" alternative, implying that a licensee would simply abandon or leave a facility after ceasing operations. This is not considered to be a viable alternative to decommissioning. The objective of decommissioning is to restore a nuclear facility to such a condition that there is no unacceptable risk from the decommissioned facility to public health and safety or the environment. In order to ensure that at the end of its life, the risk from a facility is within acceptable bounds, some action is required. If nuclear power plants were not decommissioned, they could degrade and become radiological hazards.

2 DECOMMISSIONING PROCESS

2.1 What terms or definitions are important to the understanding of decommissioning?

A number of terms are important to the understanding of decommissioning. These terms are listed and defined below. It is also important to gain an understanding of the units used for measuring radiation dose: rem and person-rem.

Activation products are radioactive materials that were created when stable substances were bombarded by neutrons. For example, cobalt-60 is formed from the neutron bombardment of the stable isotope cobalt-59. In a reactor facility, neutrons are created inside the reactor vessel during the fission process.

These neutrons bombard (1) the metal around the reactor vessel, (2) the primary reactor coolant, and (3) the concrete near the reactor vessel and create activation products in these materials.

Alpha radiation is a positively charged particle ejected spontaneously from the nuclei of some radioactive elements. It does not penetrate very far into material and it has a very short range even in air (a few centimeters). The most energetic alpha particle will generally fail to penetrate the dead layers of cells covering the skin and can be easily stopped by a sheet of paper. Alpha particles are hazardous when an alpha-emitting isotope is inside the body.

Background radiation means the radiation that is in the natural environment, including cosmic rays and radiation from the naturally radioactive elements, both outside and inside the bodies of humans and animals. It also includes radon (from the ground) and global fallout (as it exists in the environment from the testing of nuclear explosive devices or from past nuclear accidents, such as Chernobyl, that contribute to background radiation and that are not under the control of the licensee).

Contamination means undesired (for example, radioactive or hazardous) material that is (1) deposited on the surface of, or internally ingrained into, structures or equipment, or (2) mixed with another material.

Dose or radiation dose is a generic term that means absorbed dose, dose equivalent, effective dose equivalent, committed dose equivalent, committed effective dose equivalent (CEDE), or total effective dose equivalent (TEDE). In the case of *radiation dose*, it is energy absorbed per unit mass. Dose is measured in rads. The metric form of the rad is a gray (Gy) (1 rad = 0.01 gray). Radiation dose received by a person is measured in units called "rem," which incorporates the biological harm of the radiation dose based on the type of ionizing radiation. A sievert (Sv) is the metric form of the rem (1 rem = 0.01 sieverts).

Greater than Class C waste is radioactive waste that is not generally acceptable for near-surface disposal. It is waste for which form and disposal methods must be different, and in general more stringent, than those specified for Class C waste. Such waste must be disposed of in a geological repository.

Half-life is the time required for half of any quantity of identical radioactive atoms to undergo radioactive decay, so that half of the atoms in the substance are no longer emitting radiation and are no longer considered to be radioactive.

Person-rem is the sum of all the radiation dose equivalents (measured in rem) that were received by an individual or by all individuals in a population group. For example, if 1,000 people each received 1/10th of a rem (100 millirem), the corresponding population dose would be 100 person-rem. Doses to an individual are usually measured in millirem. A sievert is the metric form of the rem (1000 millirem = 1 rem = 0.01 sievert).

Radiation (ionizing radiation) means alpha particles, beta particles, gamma rays, x-rays, neutrons, high-speed electrons, high-speed protons, and other particles capable of producing ions. Radiation, as used in this section, does not include non-ionizing radiation, such as radio or microwaves, or visible, infrared, or ultraviolet light.

Radioactive decay is the spontaneous natural process by which an unstable radioactive nucleus releases energy or particles.

Rem (see *Dose*).

Residual radioactivity means radioactive contamination or activation products that remain following decontamination and dismantlement of the facility.

2.2 What is the difference between radioactive contamination and activation products, and where are contaminated materials and activated materials located?

Radioactive contamination is radioactive material that is deposited on a nonradioactive surface. The material may be deposited from the air, or it may be dissolved in water and subsequently deposited into material such as concrete. Radioactive contamination is generally located on or near the surface of materials like metal or high-density concrete or painted walls. It would travel farther into unpainted surfaces or lower density concrete. Radioactive contamination can usually be removed from surface areas by washing, scrubbing, spraying, or, in extreme cases, by removing the outer surface of the material.

Contaminated materials are transported through the facility by workers, equipment, and to some degree through the air. Although many precautions are taken to prevent the movement of contaminated material in a nuclear facility and to clean up any contaminated materials that may be found, it is most likely that contamination will occur in the reactor building, around the spent fuel pool, and around specific pieces of equipment in the auxiliary building. The areas known to contain contamination are marked by the licensee, who routinely checks for contamination.

Activation products are radioactive materials that were created when stable substances were bombarded by neutrons. Typically these materials are the concrete and the steel that surround the fuel core. The radioactive decay of activation products is the main source of radiation exposure to plant personnel.

2.3 How is a nuclear power plant decommissioned?

To decommission a nuclear power plant, the radioactive material on the site must be reduced to levels that would permit termination of the NRC license. This involves removing the spent fuel (the fuel that has been in the reactor vessel), dismantling any systems or components containing activation products (such as the reactor vessel and primary loop), and cleaning up or dismantling contaminated materials. All activated materials generally have to be removed from the facility and shipped to a waste processing, storage or disposal facility. Contaminated materials may either be cleaned of contamination onsite, or the contaminated sections may be cut off and removed (leaving most of the component intact in the facility), or they may be removed and shipped to a waste processing, storage, or disposal facility. The licensee decides how to decontaminate material, and the decision is usually based on the amount of contamination, the ease with which it can be removed, and the cost to remove the contamination versus the cost to ship the entire structure or component to a waste-disposal site.

2.4 Who decides how a facility should be decommissioned?

The licensee decides how to decommission the site. Frequently, licensees hire contractors that specialize in decommissioning sites to conduct part or most of the decommissioning. The process for decontamination and dismantlement may vary from site to site. Factors that are used to make these decisions include cost, worker exposure, availability of a waste site, and layout and structure of buildings. For example, at some sites, it may make more sense to segment the reactor vessel before removing it from the reactor building; in other cases, it would be appropriate to remove the reactor vessel intact through a hole cut in the side of the containment building and ship the reactor vessel intact.

The NRC provides a process, which is discussed in the response to Question 4.2.14, that enables the licensee to make specific types of changes without prior NRC approval. If any major decommissioning activity does not meet the conditions specified by the NRC, the licensee is prohibited from undertaking the activity until; 1): it submits a license-amendment request that describes the proposed activity and the potential impact associated with that activity, and 2) the NRC approves the request.

2.5 Is there a limit on the number of years that it would take to decommission a plant?

The NRC regulations state that decommissioning must be completed within 60 years of permanent cessation of operations. A duration of 60 years was chosen because it roughly corresponds to 10 half-lives for cobalt-60, one of the predominant isotopes remaining in the facility. By 60 years, the initial short-lived isotopes, including cobalt-60, will have decayed to background levels. In addition, the 60-year period appears to be reasonable from the standpoint of expecting institutional controls to be maintained. For periods beyond 60 years, institutional controls such as those discussed in the response to Question 8.13 would be required. Completion of decommissioning beyond 60 years will be approved by the NRC only when necessary to protect public health and safety.

2.6 What alternatives are currently used for decommissioning?

The NRC has evaluated the environmental impacts of three general methods for decommissioning power facilities.

DECON: The equipment, structures, and portions of the facility and site that contain radioactive contaminants are removed or decontaminated to a level that permits termination of the license shortly after cessation of operations.

SAFSTOR: The facility is placed in a safe stable condition and maintained in that state until it is subsequently decontaminated and dismantled to levels that permit license termination. During SAFSTOR, a facility is left intact, but the fuel has been removed from the reactor vessel, and radioactive liquids have been drained from systems and components and then processed. Radioactive decay occurs during the SAFSTOR period, thus reducing the quantity of contaminated and radioactive material that must be disposed of during decontamination and dismantlement.

ENTOMB: Radioactive structures, systems, and components are encased in a structurally long-lived substance, such as concrete. The entombed structure is appropriately maintained, and continued surveillance is carried out until the radioactivity decays to a level that permits termination of the license.

2.7 What are the benefits and costs of the DECON alternative?

The DECON option calls for prompt removal of radioactive material to permit restricted or unrestricted access. The advantages of DECON include the following:

- facility license is terminated quickly, and the facility and site become available for other purposes

- availability of the operating reactor work force that is highly knowledgeable about the facility

- elimination of the need for long-term security, maintenance, and surveillance of the facility, which would be required for the other decommissioning alternatives

- greater certainty about the availability of low-level waste facilities that would be willing to accept the low-level radioactive waste

- lower estimated costs compared to the alternative of SAFSTOR, largely as a result of future price escalation because most activities that occur during DECON would also occur during the SAFSTOR period, only at a later date. (It is anticipated that the later the date for completion of the decommissioning, the greater the cost.)

The disadvantages of DECON include the following:

- higher worker and public doses (because there is less benefit from radioactive decay such as would occur in the SAFSTOR option)

- a larger initial commitment of money

- a larger potential commitment of disposal-site space than for the SAFSTOR option

- the potential for complications if spent fuel must remain on the site until a Federal repository for spent fuel becomes available.

2.8 What are the benefits and costs of the SAFSTOR alternative?

The benefits of SAFSTOR include the following:

- a substantial reduction in radioactivity as a result of the radioactive decay that results during the storage period

- a reduction in worker dose (as compared to the DECON alternative)

- a reduction in public exposure because of fewer shipments of radioactive material to the low-level waste site (as compared to the DECON alternative)

- a potential reduction in the amount of waste disposal space required (as compared to the DECON alternative)

- lower cost during the years immediately following permanent cessation of operations

- a storage period compatible with the need to store spent fuel onsite.

Disadvantages of SAFSTOR include the following:

- shortage of personnel familiar with the facility at the time of deferred dismantlement and decontamination

- site unavailable for alternate uses during the extended storage period

- uncertainties regarding the availability and costs of low-level radioactive waste sites in the future

- continuing need for maintenance, security, and surveillance

- higher total cost for the subsequent decontamination and dismantlement period (assuming typical price escalation during the time the facility is stored).

2.9 What are the benefits and costs of the ENTOMB alternative?

The benefits of the ENTOMB process are primarily related to the reduced amount of work in encasing the facility in a structurally long-lived substance, and thus, reducing the worker dose from decontaminating and dismantling the facility. In addition, public exposure from waste transported to the low-level waste site would be minimized. The ENTOMB option may have a relatively low cost. However, because most power reactors will have radionuclides in concentrations exceeding the limits for unrestricted use even after 100 years, this option may not be feasible under the current regulations. This option might be acceptable for reactor facilities that can demonstrate that radionuclide levels will decay to levels that will allow restricted use of the site. Three small demonstration reactors have been entombed. Currently, no licensees have proposed the ENTOMB option for any of the power reactors undergoing decommissioning.

2.10 Is the choice of decommissioning alternatives a decision that is left entirely to the licensee, or does the NRC help make this decision?

The choice of the decommissioning method is left entirely to the licensee. However, the NRC would require the licensee to re-evaluate its decision if the choice (1) could not be completed as described, (2) could not be completed within 60 years of the permanent cessation of plant operations, (3) included activities that would endanger the health and safety of the public by being outside of the NRC's health and safety regulations, or (4) would result in a significant impact to the environment.

2.11 Must a licensee choose either DECON or SAFSTOR, or can it combine the two alternatives?

A licensee need not restrict its choice of decommissioning options to either an immediate decontamination and dismantlement or to a storage period of 30 to 60 years, followed by decontamination and dismantlement. Generally licensees combine the first two options. For example, after power operations stop at a facility, a licensee could use a short storage period for planning purposes, followed by removal of large components (such as the steam generators, pressurizer, and reactor vessel internals), place the facility in storage for 30 years, and eventually finish the decontamination and dismantlement process.

2.12 What main factors affect a licensee's choice of a decommissioning alternative?

The SAFSTOR alternative is often used at multi-unit sites when one or more of the units shuts down while others continue to operate. This is especially true for facilities that share some systems. In this case, the staff from the operating unit(s) assist in the maintenance and surveillance of the unit that is in storage.

The choice of decommissioning options is also strongly influenced by potential uncertainties in low-level waste disposal costs and by concerns over the future availability of low-level waste sites. The licensee's rate regulator can also influence the choice of decommissioning alternatives.

2.13 How long does the dismantlement phase last?

The dismantlement phase typically takes between 3 to 5 years to complete, although it may take longer if there are constraints on access to low-level waste burial sites, or if the licensee decides to proceed at a slower pace for programmatic reasons.

3 DECOMMISSIONED SITES

3.1 What NRC-regulated plants have been or are being decommissioned, and what decommissioning alternatives have been or are being used?

As of May 2000, radiological decommissioning had been completed at three NRC-licensed power reactors:

- Pathfinder test reactor, Sioux Falls, South Dakota

- Fort St. Vrain

- Shoreham Nuclear Power Station, Suffolk County, New York (The Shoreham plant had only operated 2 effective full-power days.)

Nine reactors are in various stages of dismantlement and decontamination:

- Saxton, Saxton, Pennsylvania

- Haddam Neck, Haddam Neck, Connecticut

- Trojan, Rainier, Oregon

- Maine Yankee, Wiscassett, Maine

- Big Rock Point, Charlevoix, Michigan

- Yankee Rowe, Franklin, Massachusetts

  Rancho Seco, Sacromento, California

  San Onofre 1, San Clemente, California

  Humboldt Bay Unit 3, Eureka, California

Ten nuclear power reactors are in, or are planning, long-term storage:

- Indian Point 1, Buchanan, New York

- Dresden 1, Morris, Illinois

- Peach Bottom, Sacramento, California

- LaCrosse, LaCrosse, Wisconsin

- Millstone 1, Waterford, Connecticut

- Vallecitos Boiling-Water Reactor (GE-VBWR), Pleasanton, California

- Fermi 1, Monroe County, Michigan

- Zion Unit 1, Zion, Illinois

  Zion Unit 2, Zion, Illinois

  Three Mile Island, Unit 2, Middletown, Pennsylvania

3.2 Have other non-NRC-regulated nuclear facilities been decommissioned?

Two nuclear power plants owned by the U.S. Department of Energy (DOE)—Elk River in Minnesota and Shippingport in Pennsylvania—have been decommissioned using the DECON alternative. Three DOE facilities—Bonus in Puerto Rico, Hallam in Nebraska, and Piqua in Ohio—have been decommissioned using the ENTOMB alternative. Other DOE test reactors and production reactors are in the process of being decommissioned.

3.3 What improvements have been made as a result of previous decommissioning experience?

Some improvements in the process, such as the removal of large components, including the reactor vessel, and the use of a primary system chemical flush to reduce worker exposure, have resulted from the experience gained from previous plant decommissionings. Other improvements or technologies were developed as part of the cleanup process for the Three Mile Island, Unit 2, plant after the 1979 accident. These include strippable coatings of latex or plastic that are used for decontaminating surfaces and the increased use of robotics. In addition, most licensees have gained experience with decommissioning techniques during routine preventive maintenance programs or as part of repairs required during operations.

3.4 What research is being performed to find improved methods to be used during decommissioning?

The U.S. Department of Energy has taken the lead on decommissioning technology research. The following types of improvements are being investigated:

- surface-removal techniques to remove the outer surface of a contaminated structure, such as lasers or microwaves combined with vacuums, electrohydraulic scabbling (water-pressure shock waves that are electrically controlled), and electrokinetic decontamination of concrete (gel electrolytes are used with electrodes to leach ionic contaminants from deep inside porous concrete)

- cutting techniques, such as laser cutting or oxy-gasoline torches (which work twice as fast as an acetylene torch on 1-inch steel) to remove structures

- improved methods for worker protection, such as protective suits with liquid air-cooling apparatus and lightweight breathable suits with chemical absorption protective layers

- environmental-protection techniques, such as automated asbestos removal and *in situ* chemical conversion of asbestos to non-hazardous material

- survey/monitoring techniques, such as pipe-explorer internal survey/characterization systems and a remote 3-D characterization and archiving system (robotic sensor and mapping platforms analyze for hazardous organic and radioactive contaminants).

Commercial firms are also developing promising avenues of research into usable technologies.

3.5 What differences are there in decommissioning between different types of reactor designs?

In 1988, the NRC issued the Final generic environmental impact statement (GEIS), NUREG-0586 (NRC 1988) a generic analysis of decommissioning at nuclear facilities, including nuclear power plants. The analysis looked at total estimated costs, occupational and public dose, and low-level waste volumes. The GEIS estimates for cost for the reference facility were generally higher by about 20% for the boiling water reactors (BWRs) than for pressurized water reactors (PWRs), depending on the decommissioning option selected. Occupational dose estimates in the GEIS were slightly higher for the reference BWR, by 10 to 50%, depending on the decommissioning option. Estimates of public dose were lower for the reference BWR by up to a factor of 2, depending upon the scenario. Burial-volume estimates for low-level waste for

10

reference BWRs and PWRs for SAFSTOR and DECON options were very close to the same. In early 2000 the NRC staff has begun an effort to update the 1988 GEIS for power reactors. The staff expects to issue a draft supplement in the year 2001.

One other major difference between decommissioning BWRs versus PWRs is that BWRs are designed so that the spent fuel pool is located in the reactor building, rather than in a separate building that can be isolated from the rest of the facility. This eliminates the possibility of decontaminating and decommissioning the remainder of the facility while leaving the spent fuel pool building as a "nuclear island."

4 NRC ACTIVITIES

4.1 Licensing

4.1.1 Does the licensee have an NRC license even during the decommissioning process?

Yes. The NRC license (called a "Part 50 license" in reference to the location of the corresponding regulations for the license in the Code of Federal Regulations) is not terminated until the licensee can demonstrate that it meets the criteria for site release in the regulations. The NRC verifies the licensee's final radiation survey by reviewing it and/or conducting a separate survey. In addition, the licensee must demonstrate that the facility has been dismantled in accordance with the approved license-termination plan.

4.1.2 Is the license the same kind that is used for an operating plant?

Yes. A facility that has permanently ceased operations and is being decommissioned maintains the same license that it had during operations. This is called a "Part 50 license," since it is regulated under Part 50 of Title 10, "Energy," of the Code of Federal Regulations. There are a number of differences between the specific regulations that apply to a decommissioning versus an operating plant. This results primarily because certain systems (for instance, the reactor coolant system) are not operating and do not need to be regulated once the fuel has been removed from the reactor vessel.

4.1.3 Could the NRC require a plant to cease operations and begin decommissioning?

The regulations allow the NRC to revoke, suspend, or modify a license in whole or in part for failure to operate a facility in accordance with the terms of the license or for violation of, or failure to observe, any of the terms and provisions of the Atomic Energy Act, regulations, license, or order of the Commission. The NRC may issue an order to a licensee to permanently cease operations. If such an order were issued, the licensee would have 60 years from the date it permanently ceased operation until the completion of decommissioning.

4.1.4 What would happen if a licensee refuses to decommission a plant that has ceased operations?

The Commission may levy a civil penalty against the licensee. The regulations allow the NRC to obtain a court order for the payment of a civil penalty for violations of any rule, regulation, or order, or for violation of any term, condition, or limitation of any license. In addition, the Atomic Energy Act provides for the Federal Government to assume responsibility for decommissioning if public health and safety are jeopardized because of inactivity on the part of the licensee.

4.1.5 What would happen if the licensee's license expires before the decommissioning process is concluded?

The decommissioning regulations state that the license for a facility that has permanently ceased operations will continue in effect beyond the expiration date to authorize possession of the facility until the Commission notifies the licensee in writing that the license is terminated. During such a period of continued effectiveness, the licensee shall take actions necessary to decommission and decontaminate the facility and shall continue to maintain the facility in a safe condition.

4.1.6 Are there any restrictions as to what the licensee can do with the site after decommissioning is completed and after the NRC has terminated the license? Would the NRC review these activities?

Frequently, after the radiological decommissioning process and the termination of the license, the licensee will remove nonradioactive facilities, or will remodel some of the remaining buildings for other industrial uses. The activities that take place after the licensee has demonstrated that the radiological hazard has been removed, and after the license has been terminated, are not within the jurisdiction of the NRC. The NRC has no further oversight of these activities once the license is terminated.

4.1.7 How would license renewal of operating plants at multi-unit facilities affect the decommissioning? Would the licensee be allowed to delay the decommissioning of the plant even longer?

The regulations specify that decommissioning will be completed within 60 years of permanent cessation of operations. Completion of decommissioning beyond the 60 years will be approved by the Commission only when necessary to protect public health and safety. Any request for such an extension would be evaluated by the Commission based on its merit. For example, a multi-unit facility with one unit that had permanently ceased operations may request and receive a license renewal for the other unit(s) that are continuing to operate. This would extend the license term for the operating unit(s) for an additional 20 years. The licensee may request an extension to the 60-year decommissioning period in order to leave the shut-down facility in a SAFSTOR mode until the operating unit(s) permanently cease operation and are ready to be decommissioned. If this occurred, the Commission would likely evaluate the condition of the shut-down unit and the positive and negative attributes of leaving the unit in SAFSTOR for a period exceeding 60 years.

4.2 Regulations

4.2.1 Who regulates the decommissioning process?

The NRC regulates and provides oversight of the radiological aspect of the decommissioning process until it agrees to terminate the license. The mission of the NRC is to ensure adequate protection of public health and safety, protection of the environment, and protection and safeguarding of nuclear materials and nuclear power plants in the interest of national security. The NRC functions include (1) licensing nuclear facilities, (2) developing regulations and regulatory guidance, (3) conducting inspections and enforcement activities to ensure compliance with the regulations, (4) reviewing changes to the license, changes to licensee programs, and unreviewed safety questions, and (5) providing licensees with information that would improve their performance.

4.2.2 Does the U.S. Environmental Protection Agency (EPA) have authority over the decommissioning process?

The NRC oversees and regulates the decommissioning process in the same way that it regulates the operation of the nuclear power plant. However, licensees are required to comply with EPA regulations related to liquid effluent discharges to bodies of water. Compliance with the Clean Water Act implementing regulations requires the licensee to obtain a National Pollutant Discharge Elimination System (NPDES) permit for routine discharges to any body of water. In many cases, the EPA may delegate the authority to write and monitor the use of NPDES permits to States; for most facilities, the NPDES permit is issued by the State where they are located. In some States, the authority still remains with EPA. Likewise, the EPA has implemented regulations for discharges to the atmosphere (Clean Air Act), and permits are required for specific pollutants, although reactor decommissioning does not generally fall under these regulations.

4.2.3 When were the current regulations for decommissioning written?

The NRC continually reviews the decommissioning regulations that it is using and revises them as required to improve the regulatory process. On July 29, 1996, a final rule amending the regulations on decommissioning procedures was published in the *Federal Register* (61 FR 39278). On July 21, 1997, the NRC published (also in the *Federal Register*) a final rule entitled, "Radiological Criteria for License Termination" (62 FR 39058). This rule contains the regulations regarding the radiological criteria that the licensee must meet before the license can be terminated. These regulations superseded regulations written in 1988. Certain financial-assurance aspects of the regulations were amended by a final rule that was published in the *Federal Register* (63 FR 50465) on September 22, 1998.

4.2.4 What publications contain the regulations for decommissioning?

Regulations regarding the decommissioning of NRC-licensed plants appear in the *Code of Federal Regulations*. The *Code of Federal Regulations* is a codification of the general and permanent rules published in the *Federal Register* by the executive departments and agencies of the Federal Government. The *Code* is divided into 50 titles, which represent broad areas subject to Federal regulation. Each title is divided into chapters; these usually bear the name of the issuing agency. Each chapter is further subdivided into parts covering specific regulatory areas.

The regulations related to the decommissioning of power reactors are included in Title 10, "Energy," Chapter I—Nuclear Regulatory Commission; for example, Part 20, "Standards for Protection Against Radiation"; Part 50—"Domestic Licensing of Production and Utilization Facilities"; and Part 51—"Environmental Protection Regulations for Domestic Licensing and Related Regulatory Functions." The subparts related to decommissioning are 20.1402, "Radiological criteria for unrestricted use"; 20.1403, "Criteria for license termination under restricted conditions"; 20.1404, "Alternate criteria for license termination"; 20.1405, "Public notification and public participation"; 20.1406, "Minimization of contamination"; 50.75, "Reporting and recordkeeping for decommissioning planning"; 50.82, "Termination of license"; 51.53, "Post-construction environmental reports"; and 51.95, "Post-construction environmental impact statements." These regulations state the technical and financial criteria for decommissioning licensed nuclear facilities. They address decommissioning, planning needs, timing, funding methods, and environmental-review requirements.

See the response to Question 13.6 for instructions on how to obtain a copy of the regulation.

4.2.5 What regulatory actions are required to decommission a nuclear facility?

The regulations specify actions that both the NRC and the licensee must take to decommission a nuclear power plant. Once the decision is made to permanently cease operations, the licensee must notify the NRC, in writing, within 30 days. The notification must contain the date on which the power generation operations ceased or will cease. The licensee must remove the fuel from the reactor and submit a written certification to the NRC confirming its action. There is no time limit specified before the fuel must be removed or the corresponding certification received by the NRC. Once this certification has been submitted, the licensee is no longer permitted to operate the reactor or to put fuel back into the reactor vessel. This also reduces the licensee's annual license fee to the NRC and eliminates the obligation to adhere to certain requirements that are needed only during reactor operations. The licensee must submit a post-shutdown decommissioning activities report (PSDAR) to the NRC and the affected State(s) no later than 2 years after the date of permanent cessation of operations. The PSDAR must

- describe the planned decommissioning activities

- contain a schedule for the accomplishment of significant milestones

- provide an estimate of expected cost

- provide documentation that environmental impacts associated with site-specific decommissioning activities have been considered in previously approved environmental impact statements.

If the environmental impacts that are identified have not been considered in existing environmental assessments, the licensee must address the impacts in a request for a license amendment regarding the activities. The licensee also must submit a supplement to its environmental report that relates to the additional impacts. The NRC will review this environmental assessment or supplement to the environmental statement in conjunction with its review of the license-amendment request.

After receiving a PSDAR, the NRC publishes a notice of receipt, makes the PSDAR available for public review and comment, and holds a public meeting in the vicinity of the plant to discuss the licensee's plans.

Although the NRC will determine if the information is consistent with the regulations, NRC approval of the PSDAR is not required. However, should the NRC determine that the informational requirements of the regulations are not met in the PSDAR, the NRC will inform the licensee in writing of the deficiencies and require that they be addressed before the licensee initiates any major decommissioning activities.

Upon completion of the required submittals, and allowing for a 90-day waiting period after submittal of the PSDAR, the licensee may commence major decommissioning activities. Major decommissioning activities include the following:

- permanent removal of major radioactive components, such as the reactor vessel, steam generators, or other components that are comparably radioactive

- permanent changes to the containment structure

- dismantling components resulting in "greater than Class C" waste.

Decommissioning activities conducted without specific prior NRC approval must not preclude release of the site for possible unrestricted use, must not result in there being no reasonable assurance that adequate funds will be available for decommissioning, and must not cause any significant environmental impact not previously reviewed. If any decommissioning activity does not meet these terms, the licensee is required to submit a license-amendment request before conducting the activity; this would provide an opportunity for a public hearing.

Activities that are not considered to be "major decommissioning activities" may be performed in accordance with the license and technical specifications in effect for the facility even before the end of the 90-day waiting period. Allowable activities would include such routine items as maintenance and low-level waste disposal of small radioactive components.

Within 2 years following the date of permanent cessation of operations, the licensees must submit a site-specific cost estimate for the decommissioning project. The licensee is prohibited from using the full amount of money that was accumulated during operations for the decommissioning process until the site-specific cost estimate is submitted to the NRC.

Within 2 years following permanent cessation of operation licenses are also required to submit for review and preliminary approval a description of how they intend to manage and provide funding for the management of irradiated fuel until title to the fuel is transferred to the Department of Energy (DOT).

Unless the licensee receives permission to the contrary, the site must be decommissioned within 60 years. The licensee remains accountable to the NRC until decommissioning has been completed and the license is terminated. In order to conclude its obligations, the licensee must submit a license-termination plan.

The license-termination plan must be submitted at least 2 years before the termination date. It must include the following:

- a site characterization

- identification of remaining dismantlement activities

- plans for site remediation

- detailed plans for the final survey of residual contamination on the site

- a description of the end-use of the site (if restricted use is proposed, a description of institutional controls and maintenance and surveillance programs is needed)

- an updated site-specific estimate of remaining decommissioning costs

- a supplement to the environmental report.

After receiving the license-termination plan, the NRC will place a notice of receipt of the plan in the *Federal Register* and will make the plan available to the public for comment. The NRC will schedule a public meeting near the facility to discuss the plan's contents with the public. The NRC will also offer an opportunity for a public hearing on the license amendment associated with the licensee-termination plan. If the license-termination plan demonstrates that the remainder of decommissioning activities will be performed in accordance with the NRC's regulations, is not detrimental to the health and safety of the public, and does not have a significant effect on the quality of the environment, the Commission will approve the plan by a license amendment (subject to whatever conditions and limitations the NRC deems appropriate and necessary). Once the license amendment is granted, the licensee may achieve license termination by successfully demonstrating implementation of the license-termination plan as approved by the NRC.

At the end of the license-termination-plan process, if the NRC determines that the remaining dismantlement has been performed in accordance with the approved license-termination plan, and if the final radiation survey and associated documentation demonstrate that the facility and site are suitable for release, then the Commission will terminate the license, and the decommissioning process is considered complete.

4.2.6 Why isn't an environmental impact statement (EIS), or environmental assessment (EA) required at the start of the decommissioning process?

In 1996, the NRC published the final rule on the current decommissioning regulations. The Commission stated in that rule that they had determined that under the National Environmental Policy Act of 1969, as amended, and under the Commission's own regulations, that major decommissioning (dismantlement) activities could proceed without an environmental assessment. The impacts of these major decommissioning activities were determined to be within the bounds of those that were assessed in the NRC's Generic Environmental Impact Statement (GEIS) on Decommissioning (NUREG-0586), which was issued in 1988. However, because the GEIS did not address site-specific situations, the final rule prohibits major decommissioning activities that could result in significant environmental impacts not previously reviewed. The review process also includes a requirement that the licensee provide the reasons for concluding that the planned activities are bounded by the GEIS and previous site-specific environmental impact statements. The license termination plan (LTP), submitted by the licensee near the end of the decommissioning process, is reviewed and approved by the NRC staff. The LTP is incorporated into the facility license by an amendment to the license. The amendment process includes an environmental assessment by the staff.

4.2.7 What activities can take place before submitting the Post-shutdown Decommissioning Activities Report?

No major decommissioning activities may take place until 90 days after the PSDAR has been submitted. Major decommissioning activities are defined as "any activity that results in permanent removal of major radioactive components, permanently modifies the structure of the containment, or results in dismantling components for shipment containing greater than Class C waste." Major radioactive components are defined by the regulations as the reactor vessel and internals, steam generators, pressurizers, large-bore reactor-coolant-system piping, and other large components that are radioactive to a comparable degree. Examples of activities considered minor decommissioning activities are (1) normal maintenance and repair, (2) removal of certain, relatively small, radioactive components, such as control-rod drive mechanisms, control rods, pumps, piping, and valves, (3) removal of components (other than those defined above as major components) similar to those removed for maintenance and repair during plant operations, (4) removal of non-radioactive components and of radiation structures not required for safety, (5) shipment of reactor fuel offsite, and (6) site characterization and measurement of contamination levels.

4.2.8 Can the licensee make changes to its plans or commitments after submitting the PSDAR?

Yes. However, the regulations require the licensee to notify the NRC in writing before performing any decommissioning activity that is not consistent with, or could be considered to be a significant change from, the actions or schedules described in the Post-Shutdown Decommissioning Activities Report (PSDAR). Significant changes in cost are also to be reported to the NRC. Significant changes to the milestone schedule are used by the NRC staff to reschedule NRC inspections of the licensee's activities. Examples of changes in activities and schedule that would require prior NRC notification include, but are not limited to, changing from long-term storage to active dismantlement, changing the method used to remove the reactor vessel from cutting and segmenting to intact removal, or changing the schedule to affect major milestones. Examples of significant increases in cost associated with decommissioning the facility would include a new estimated cost more than 20 percent above the site-specific cost estimate or the PSDAR cost estimate, or a 25-percent increase in cost above a major milestone estimate.

Written notifications to the NRC do not require a 90-day waiting period before initiation of activities. Typically, the staff would not require a public meeting to discuss the proposed changes unless the NRC staff determined that the change was significant enough to warrant a public meeting.

4.2.9 What would happen if the licensee wanted to switch from SAFSTOR to DECON?

If the licensee proposes changing the method of decommissioning, for example, from long-term storage followed by decontamination and dismantlement to prompt decontamination and dismantlement, a public meeting would be held. Normally, a licensee's decision to conduct some limited decontamination and dismantlement during long-term storage (if such action had not been specified in the PSDAR) would require submittal of an update, but would not require a public meeting. If the expected environmental impact of any change in decommissioning activities is significantly greater than that predicted in a previous environmental impact statement or assessment (see NUREG-0586, "Final Generic Environmental Impact Statement on Decommissioning of Nuclear Facilities"), the licensee would be required to request a licensee amendment and provide a supplement to the environmental report for the facility that evaluates the impact of the change. The NRC staff would review the licensee's submittal and would publish either an environmental assessment or a supplement to the facility's final environment statement in conjunction with review of the license-amendment request.

4.2.10 Will the PSDAR or the License Termination Plan be reviewed by the U.S. Environmental Protection Agency (EPA)?

The NRC will place a notice in the *Federal Register* of the availability of the PSDAR for comment. The EPA, as all Federal agencies, will be invited to comment on the PSDAR.

4.2.11 Under what circumstances would the NRC refuse to allow a licensee to proceed with the decommissioning?

It is highly unlikely that the NRC would refuse to allow a licensee to decommission a plant since the regulations require a licensee to complete the decommissioning process within 60 years of permanent cessation of operations. However, the NRC staff may determine that the information in the PSDAR is inconsistent with what is required by the regulations. In that case, the NRC will inform the licensee in writing of the deficiencies and will require the deficiencies to be addressed before starting major decommissioning activities.

4.2.12 Would the NRC ever stop a licensee from decommissioning a plant on the basis of information in the Post-Shutdown Decommissioning Activities Report?

A number of factors could cause the NRC to find the PSDAR deficient, and the NRC staff would prevent the licensee from proceeding with decommissioning as described in the PSDAR. The NRC could find the PSDAR deficient if the licensee's plan for decommissioning could not be completed as described (for example, if the plan called for an immediate decontamination and dismantlement of the facility and there were no waste-disposal facilities available for the licensee to use). The NRC could find the PSDAR deficient if the schedule contained a decommissioning process that could not be completed within 60 years of the permanent cessation of operations, unless it were shown that this action was necessary to protect public health and safety. Factors that would be considered for an extended decommissioning process include the unavailability of low-level-waste disposal capacity and other site-specific factors affecting the licensee's capability to carry out decommissioning in the given time period, including the presence of other operating nuclear facilities at the site. The NRC would find the PSDAR deficient if the licensee's decommissioning plans, as presented in the PSDAR, contained a decommissioning process that obviously could not be completed for the estimated cost (the NRC staff will base this decision on the generic guidelines and on previous facility decommissioning costs). The NRC would also find a PSDAR deficient if it contained activities that would endanger the health and safety of the public by being outside the NRC's health and safety regulations or that would result in a major detrimental impact to the environment that was not bounded by the current environmental impact statements.

4.2.13 Can the NRC require the licensee to follow the regulations, and how would the NRC know if the regulations were not being followed?

The NRC reviews licensee procedures and works to ensure that licensees are properly adhering to the regulations. At the start of the decommissioning process, the NRC reviews the licensee's plans to enter active decontamination and dismantlement. Generally, an NRC resident inspector will remain onsite for a time after cessation of operations. Additionally, all through the decommissioning process, NRC inspectors will periodically conduct special inspections of specific activities at the site. Site visits and inspections will be more frequent for plants that are undergoing decontamination and dismantlement, and less frequent for plants that are in a storage mode. If the licensee is not complying with regulations, the NRC will address the issue in accordance with NRC's enforcement policy (NUREG-1600).

4.2.14 Does the NRC have to review and either approve or disapprove every decision the licensee makes related to decommissioning a facility?

No. The NRC does not review every decision the licensee makes related to decommissioning the facility. The licensee may make any of the following types of changes without prior NRC approval:

- changes in the facility as described in the safety analysis report

- changes in the procedures as described in the safety analysis report

- conduct of tests or experiments not described in the safety analysis report.

These changes, tests, and experiments can be made or conducted without prior Commission approval provided that the proposed change, test, or experiment does not require a change (1) in the technical specifications incorporated in the license or does not meet any of the following criteria: (1) that the likelihood of occurrence or the consequences of an accident or malfunction of structures, systems or equipments important to safety and previously evaluated in the safety analysis report may be increased more than a minimal amount, or (2) that the possibility of an accident or malfunction of a different type from any evaluated previously in the safety analysis report (SAR) may be created, or (3) create a possibility for a malfunction of a structure, system or component important to safety with a different result than any previously evaluated in the SAR, or (4) result in a design basis limit for a fission product barrier as described in the SAR. being exceeded or (5) result in a departure from a method of evaluation described in the SAR used in establishing the design basis or in the safety analysis. These revised requirements are contained in 10 CFR 50.59 and will be in effect January 2001.

4.2.15 How can the NRC make sure a licensee does not make a mistake in its evaluation of proposed changes, tests, or experiments? What happens if the licensee fails to identify an unreviewed safety question?

The licensee is required to periodically submit a report containing a brief description of any changes, tests, and experiments made at the facility, and to summarize the safety evaluation of each. This report must be filed with the NRC at least every 24 months for facilities that have submitted certifications for permanent removal of fuel. These reports are reviewed by the NRC. In addition, records of changes in the facility are required to be maintained until the date of termination of the license, and records of changes in procedures and records of tests and experiments are required to be maintained for a period of 5 years. These changes in the facility are normally inspected by the NRC annually. The Decommissioning Power Reactor Inspection Program Manual specifies the review of the documents that detail the licensee's basis for conducting operations in a safe manner. If the inspector disagrees with the licensee's evaluation or if there are questions as to whether a proposed change is an unreviewed safety question, the NRC staff will evaluate this issue until it is appropriately resolved.

4.2.16 What would happen if the licensee decides to do something that will endanger the environment during the decommissioning process?

NRC regulations state that the licensee must not perform any decommissioning activity that, (1) forecloses release of the site for possible unrestricted use, (2) causes any significant environmental impact not previously reviewed, or (3) results in there no longer being reasonable assurance that adequate funds will be available for decommissioning. If any major decommissioning activity could not meet these conditions, the licensee is prohibited from undertaking the activity until it submits a license-amendment request that describes the proposed activity and the potential impact associated with that activity. The license-amendment request would provide an opportunity for a public hearing. The NRC staff will evaluate the licensee's procedures for ensuring that these three restrictions are part of the screening criteria that are used for proposed changes to the facility. Additionally, periodic inspections of the licensee's activities will consider the environmental protection issues.

4.2.17 What were the previous regulations and why were they changed?

When the NRC issued decommissioning regulations in 1988, it was assumed that decommissioning would take place after the facility's operating license expired. The licensee was obligated to submit a preliminary decommissioning plan 5 years before the license expired. The preliminary decommissioning plan contained a cost estimate for decommissioning and an up-to-date technical assessment of the factors that could affect planning for decommissioning. This included (1) the choice of alternative, (2) the major technical actions necessary to carry out decommissioning safely, (3) the current situation with regard to disposal of high-level and low-level radioactive waste, (4) the residual radioactivity criteria, and (5) other site-specific factors that could affect decommissioning planning and cost.

The previous rule also required that no later than 1 year before expiration of the license (or within 2 years of permanent cessation of operations for plants closing before their license expires), a licensee had to submit an application for authority to decommission the facility. The application was to be accompanied by or preceded by a proposed decommissioning plan. The proposed decommissioning plan was to include (1) the choice of the alternative for decommissioning with a description of the activities involved, (2) a description of controls and limits on procedures and equipment to protect occupational and public health and safety, (3) a description of the planned final radiation survey, (4) an updated cost estimate for the chosen alternative and a plan for ensuring the availability of adequate funding, and (5) a description of the technical specifications, quality assurance provisions, and physical security plan provisions in place during decommissioning. A supplemental environmental report that described any substantive environmental impacts that were anticipated but not already covered in other environmental impact documents was also required.

The NRC reviewed the decommissioning plan. The Commission would approve the plan if the plan demonstrated that the decommissioning would be performed in accordance with regulations and there were no security, health, or safety issues. The Commission would also require that notice be given to interested persons. However, the NRC could add other conditions and limits to the plan that it deemed appropriate. The license would then be terminated if the Commission determined that the decommissioning had been performed in accordance with the approved decommissioning plan and the order authorizing decommissioning, and if a final radiation survey and associated documentation demonstrated that the facility and site were suitable for release for unrestricted use.

The regulations were revised for several reasons. First, the experience gained in the early decommissioning activities associated with several facilities did not reveal any activities that required NRC review and approval of a decommissioning plan. Second, environmental impacts associated with decommissioning those early facilities resulted in impacts consistent with those evaluated in the "Generic Environmental Impact Statement on Decommissioning of Nuclear Facilities," NUREG-0586. And finally, experience gained from reviewing numerous decommissioning oversight activities at a number of these facilities also indicated that the decommissioning activities were in general no more complicated than activities normally undertaken at operating reactors without prior and specific NRC approval. In August 1996, the revised rule that redefined the decommissioning process and required licensees to provide the NRC with early notification of planned decommissioning activities at their facilities went into effect. The rule made the decommissioning process more efficient and uniform. It provided for greater public participation in the decommissioning process and gave plant personnel a clearer understanding of the process for changing from an operating organization to a decommissioning organization. The current regulatory process for decommissioning a facility is described in the response to Question 4.2.5.

4.3 Inspection Program

4.3.1 What are the goals of the inspection program at nuclear power plants undergoing decommissioning?

The goals of the inspection program at nuclear power plants undergoing decommissioning are to

- obtain sufficient information through direct observation and verification to determine if decommissioning is being conducted safely, if the spent nuclear fuel is being stored safely, and if activities at the site are being conducted in accordance with all applicable regulations and commitments

- determine if the administrative controls that the licensee has in place are adequate and in accordance with regulatory requirements (the controls include self-assessment, audits and corrective actions, design control, safety review, maintenance and surveillance, radiation protection, and effluent controls)

- identify any significant declining performance trends and verify that the licensee has taken actions to reverse any trend.

4.3.2 What types of inspections are conducted?

Three types of inspections will be conducted: core, temporary instruction compliance, and discretionary.

Core inspections involve the inspection of a number of functional areas at specified frequencies. The functional areas are facility management and control, decommissioning support activities, spent fuel safety, and radiological safety. Each of these functional areas is further broken down into a number of subelements. Each subelement has a specific frequency based on whether the plant is actively removing components, is in a storage mode, or is undergoing site-termination activities.

Temporary instruction compliance inspections are inspections designed to investigate generic problems that have been identified. Examples are spent fuel pool siphoning and spent fuel pool seismic protection.

Discretionary inspections are detailed reviews of a particular functional area. Examples are high-level waste transportation, internal and external dose assessment, and confirmatory surveys.

4.3.3 Which areas receive the greatest emphasis during the inspections?

The inspection effort during decommissioning places the greatest emphasis on radiological controls, management, procedure compliance, spent fuel pool operations, and the safety review program.

4.3.4 Are there some decommissioning activities that receive greater oversight than others?

Many of the activities that occur during decommissioning are very routine and occur frequently in operating plants. These activities include decontamination of surfaces and components, and waste packaging and disposal, among others. The NRC inspectors have previously reviewed the process the licensee uses for many of these activities and will continue to review the process as any changes are made. The types of activities that will receive the most attention from the NRC inspectors include any activities that relate to the safe storage of irradiated fuel and any activities that could result in offsite releases in excess of the regulatory limits (for example, removal of a large component like a steam generator).

4.3.5 Does the NRC independently monitor radioactive releases into the environment?

The NRC does not independently monitor radioactive releases on a continuous basis. However, the NRC does occasionally have an independent measurement of the releases to verify that the licensee's program is adequate to continuously analyze samples and to measure the amount of radioactive material that leaves the plant. In addition, the NRC reviews the licensee's documented procedures for monitoring radioactive releases into the environment and reviews the results periodically. All licensee's submit effluent release results to the NRC on an annual basis.

4.3.6 Will the NRC have inspectors onsite during decommissioning?

The inspection effort at a plant undergoing decommissioning is significantly less than at an operating reactor site. Operating reactor facilities have between one and two onsite inspectors, who maintain a continual presence at the plant or in the vicinity, for each unit on the site. However, because of the reduced hazard during the decommissioning process, NRC inspectors may not maintain a continual presence. The NRC will generally remove the onsite inspector from a single-unit site within 1 year from a reactor's permanent cessation of operations. The resident inspector's stay at a site can be extended because of special circumstances, however. Rather than stationing a resident inspector at the site during the entire decommissioning process, the NRC will provide subject-matter experts to cover specific activities occurring at the site. For example, if the licensee is planning to remove a large component, the NRC might send at appropriate times an expert in radiation protection, an expert in heavy lifting and polar cranes, and an expert in packaging radioactive waste. Inspections are performed by the NRC headquarters staff and NRC regional personnel. This requires attention to scheduling so that NRC personnel are available to review the licensee's procedures and to inspect before and during specific activities. The extent of onsite presence at the facility will depend on the activities that are taking place. During active decommissioning, NRC personnel may be at the facility 2 or 3 weeks of the month. During storage operations, they would be present several times a year.

4.3.7 Are inspection reports associated with the decommissioning activities available on the NRC's website?

The inspection manual chapters are available from the NRC website (http://www.nrc.gov/NRC/IM/index.html). The inspection reports are available from the Public Document Room and from the Electronic Reading Room. Section 13 describes the process for obtaining information from the Public Document Room and the Electronic Reading Room.

4.3.8 Will plant security levels be de-emphasized during decommissioning?

Licensees have requested and been granted exemptions to specific sections of 10 CFR 73.55 that would reduce the security requirements below those of an operating reactor. This is done only after a thorough review by NRC security specialist.

5 SPENT FUEL

5.1 What are "high-level wastes?"

High-level radioactive waste (HLW), as it pertains to commercial nuclear power reactors, is mainly irradiated (spent) reactor fuel.

5.2 What is meant by the term "spent fuel?"

Spent nuclear fuel is uranium-bearing fuel elements that have been used at commercial nuclear power reactors. Although spent (used) fuel can no longer produce enough heat to produce electricity, it contains highly radioactive material resulting from the fission process that takes place within the reactor. As a result, it still continues to generate radiation and heat. This heat and radiation are caused by "radioactive decay" of the products of the fission process. The heat and radioactivity in spent fuel necessitate that any shipment be made in containers or casks that provide the necessary degree of protection. In practice, this means that a cask must shield and contain the radioactivity and dissipate the generated heat.

5.3 Are there facilities or plans for facilities for the disposal of high-level radioactive waste?

At this time, there are no facilities for permanent disposal of high-level radioactive wastes. On January 7, 1983, the President signed into law the Nuclear Waste Policy Act, which defined the goals and structure of a program for permanent, deep geologic repositories for the disposal of high-level radioactive waste and unreprocessed spent fuel. Under this Act, the U.S. Department of Energy (DOE) is responsible for developing permanent disposal capacity for the spent fuel and other high-level nuclear wastes. At the present time, DOE, as directed by Congress, is investigating a site in Yucca Mountain, Nevada, which would be built and operated by DOE and licensed by the NRC, for a possible disposal facility.

5.4 Is the licensee allowed to store the spent fuel in the reactor vessel?

No. The licensee must submit a certification indicating that it has permanently removed the fuel from the reactor vessel before it can start the decommissioning process. When the NRC receives this certification, the licensee is prohibited from loading fuel back into the reactor vessel. The licensee has incentives to permanently remove the fuel from the reactor vessel because certification (along with the certification of permanent cessation of operations) reduces the licensee's annual license fee to the NRC and eliminates the obligation to adhere to certain requirements needed only during reactor operations.

5.5 Since there are no facilities for permanent disposal, where will the spent nuclear fuel be kept during the decommissioning process?

Until the repository is approved and constructed, spent nuclear fuel is being stored primarily in specially designed, water-filled basins (spent fuel pools) at individual reactor sites around the country. Another option for storage is in an independent spent fuel storage installation (ISFSI), which is located either at the reactor site or elsewhere. The spent fuel may be stored in air-cooled dry casks at an ISFSI (see Section 5.9). These options will continue to be available during the decommissioning process.

5.6 What are the long-range plans for disposition of spent fuel?

The fuel would be stored in an independent spent-fuel storage installation (ISFSI) or in the spent fuel pool until a Federal repository became available. Then the spent fuel would be shipped to the Federal repository for final disposition.

5.7 What happens if a disposal site for high-level waste is never licensed?

The NRC has stated in its regulations that "The Commission has made a generic determination that, if necessary, spent fuel generated in any reactor can be stored safely and without significant environmental impact for at least 30 years beyond the licensed life for operation (which may include the term of renewed license) of that reactor at its spent fuel storage basin or at either onsite or offsite independent fuel-storage installations." Further, the Commission believes there is reasonable assurance that at least one mined geological repository will be available in the first quarter of the 21st century, and sufficient repository capacity will be available within 30 years beyond the licensed life for operation of any reactor to dispose of the commercial high-level waste and spent fuel originating in such reactor and generated up to that time.

5.8 Spent Fuel Pools

5.8.1 Why is spent fuel stored in a pool of water?

Even after the nuclear reactor is shut down, the fuel continues to generate decay heat. Decay heat results from the radioactive decay of fission products. The rate at which the decay heat is generated decreases the longer the reactor has been shut down. So the longer the spent fuel has been out of the reactor, the less heat that it gives off. Storing the spent fuel in a pool of water is a way to provide an adequate heat sink for the removal of heat from the irradiated fuel. In addition, the fuel is located far enough under water that the radiation emanating from the fuel is shielded by the water to adequately protect the workers from the radiation.

5.8.2 Has the spent fuel pool been analyzed to determine the limits for heat removal due to spent fuel storage?

Yes. The regulations give criteria that must be met for fuel storage and handling. This includes designing fuel-storage systems to ensure adequate safety under normal and postulated accident conditions. The system is to be designed with suitable shielding for radiation protection, with appropriate containment, confinement, and filtering systems, and with a heat-removal capability that is reliable and that can be tested to ensure that it meets the requirements for removing the heat produced by the spent fuel.

5.8.3 Do spent fuel pools leak, and if they do, how much radioactive material could be leaked, and where would it go?

All nuclear power plants have a reinforced-concrete spent fuel pool (SFP) structure designed to retain its function, even following the design-basis seismic event (that is, seismic Category 1 or Class 1 [earthquake]) that is anticipated for the area. The SFP also has a welded, corrosion-resistant liner. All plants except for one have leak-detection channels positioned behind liner plate welds to collect any leakage and to direct the leakage to a point at which it can easily be monitored. Nearly all nuclear power reactors have passive features preventing draining or siphoning of the SFP to a coolant level below the top of stored, irradiated fuel. Excluding paths used for irradiated fuel transfer, passive features at nearly all nuclear reactors prevent draining or siphoning the coolant to a level that provides inadequate shielding for fuel seated in the storage racks.

In the event that SFP coolant inventory decreases significantly, several indicators are available to alert operators to that condition. The primary indication is a low-level alarm. A secondary indication of a loss of coolant is provided by area radiation alarms. These primary and secondary alarms indicate a loss of shielding that occurs when SFP coolant inventory is lost. Except for the SFP located inside the containment building, the area radiation alarms are set to alarm at a level low enough to detect a loss of coolant inventory early enough to allow for recovery before radiation levels could make such a recovery difficult.

The level of radioactivity in the water of the spent fuel pool is low. In addition, nuclear plants have cleaning systems to maintain the purity of the water in the spent fuel pools. All nuclear plants have a groundwater monitoring system around the facility so that if a system leaks, there is a method for alerting the licensee to the problem as well as for providing information regarding the location of the contamination.

5.8.4 What would happen if there were a loss of heat-removal capability of water in a spent fuel pool when it was fully loaded?

The consequences of losing the heat-removal capability or water (coolant) in a spent fuel pool depends on the amount of time since the fuel was last used for power operation inside the reactor. If fuel was recently used for power operation, there may be enough decay heat to cause the spent fuel pool coolant to heat up to the boiling point if forced cooling were lost to the spent fuel pool. If plant operators took no action, boiling would cause the level in the spent fuel pool to decrease over time. However, operators have redundant sources of water to add to the pool to maintain coolant level should a loss of forced cooling occur. Operators are alerted to a loss of level condition by a series of alarms at the cooling system control station and in the main control room. Given the unlikely event that no operator action is taken, the pool level would decrease at a very slow rate (about one foot every several hours to weeks, depending on the age of the stored fuel). The longer the time interval since the last batch of fuel was used to generate power in the reactor, the longer it would take for the spent fuel to boil off the spent fuel pool coolant. Boiling the spent fuel pool coolant is, however, an acceptable method for cooling spent fuel and has a minimal effect on public health and safety. In the unlikely event that a large loss of coolant uncovers the spent fuel, if sufficient time has not elapsed since the fuel was used to generate power in the reactor, the spent fuel may have enough decay heat to overheat the cladding in air and cause it to ignite. The resulting fire could carry radioactive particles offsite and the consequences could be significant. However, the NRC staff considers this a very low probability accident because of design features required at all spent fuel storage pools that minimize the possibility of losing all of the spent fuel pool coolant.

5.8.5 What would happen to the fuel in the spent fuel pool if an earthquake ruptured the pool, or if an airplane crashed into the pool?

Spent fuel pools are designed to withstand earthquakes greater than any earthquake that actually occurred or is expected to occur in the area of the plant. Therefore, the probability of the spent fuel pool rupturing due to an earthquake is very low. However, in the unlikely event that a very large earthquake does occur, one that is larger than the pool was designed to withstand, the pool structure could fail and allow the coolant to drain out. The consequences of an accident like this are discussed in the response to Question 5.8.4, above. In the unlikely event that an aircraft crashed into the spent fuel pool, the pool structure could be severely damaged and not capable of maintaining coolant level. In this event, consequences such as those discussed in Question 5.8.4 could result. However, the staff has evaluated the possibility of an aircraft impacting the spent fuel pool and consider it a very low probability event.

5.8.6 What can be done to prevent the spent fuel pool from boiling dry?

A cooling system removes decay heat from the spent fuel pool. The coolant in the spent fuel pool is maintained below a specific temperature and the level of the water is maintained at a specific height over the spent fuel. Temperature indicators are installed and are either equipped with an alarm or require visual surveillance on a daily basis. High/low-water-level monitors are installed in spent fuel pools. The monitor alarms at the spent fuel pool and in the control room when the spent fuel pool's water level is not within the specified limit.

5.8.7 How long can the licensee store the spent fuel in the spent fuel pool?

The Commission has made a generic determination in 10 CFR 51.23 that, if necessary, spent fuel generated in any reactor can be stored safely and without significant impacts for at least 30 years beyond the licensed life for operation.

5.8.8 What will be done with the spent fuel pool after the fuel has been removed?

The spent fuel pool will be decontaminated and most likely dismantled.

5.8.9 Could the licensee transfer fuel to another licensee's facility?

If the fuel that came out of the reactor vessel were not fully expended (had not been thoroughly used or "burned up"), it could be shipped to another licensee's facility for use there. This would offset some of the licensee's costs. Spent fuel that has no further value in producing power could also be shipped to other facilities. Amendment to one or both of the facility licenses would be required before fuel transfer. The fuel would have to be transported in a licensed cask that met NRC requirements for shipping.

5.8.10 Can the spent fuel be shipped to another facility's spent fuel pool for storage?

Yes. NRC regulations do not prohibit spent fuel from one facility being stored in the spent fuel pool at another facility. As noted in the response to Question 5.8.9, amendment to one or both of the facility licenses would be required before shipment. The spent fuel would have to be transported in a licensed cask that met NRC requirements for shipping.

5.8.11 Where can I find the regulations relating to the storage of spent fuel in a spent fuel pool?

Regulations regarding the storage of fuel in a spent fuel pool appear in the *Code of Federal Regulations*. The *Code of Federal Regulations* is a codification of the general and permanent rules published in the *Federal Register* by the executive departments and agencies of the Federal Government. The *Code* is divided into 50 titles, which represent broad areas subject to Federal regulation. Each title is divided into chapters; these usually bear the name of the issuing agency. Each chapter is further subdivided into parts covering specific regulatory areas.

The regulations related to spent fuel pools are in Title 10, "Energy," Chapter I – Nuclear Regulatory Commission, Part 50, "Domestic Licensing of Production and Utilization Facilities," along with the regulations and standards for construction permits and operating licenses for nuclear power plants.

See the response to Question 13.6 for instructions on how to obtain a copy of the regulations.

5.9 Independent Spent Fuel Storage Installation

5.9.1 What is an independent spent fuel storage installation (ISFSI)?

An independent spent fuel storage installation or ISFSI is a complex designed and constructed for the interim storage of spent nuclear fuel and other radioactive materials associated with spent fuel storage. ISFSIs may be located at the site of a nuclear power plant or at another location. The most common design for an ISFSI at this time is a concrete pad with dry casks containing spent fuel bundles.

5.9.2 Why would a licensee store spent fuel in an ISFSI rather than in the spent fuel pool?

ISFSIs are used by operating plants that require increased spent fuel storage capability because their spent fuel pools have reached their capacity for holding spent fuel. Decommissioning facilities also use ISFSIs. Licensees that remove the spent fuel from their pools and place it in an ISFSI can then continue and/or complete the decommissioning process on the power generation facilities and subsequently terminate the facility license. In some instances, the license for the nuclear power reactor can be terminated when the ISFSI, which has a separate license, is located on the facility site.

In general, it is both less expensive and easier to maintain spent fuel in dry storage in an ISFSI than in a spent fuel pool. The cost of an ISFSI as compared to spent fuel pool storage depends on many factors unique to each facility. Dry cask storage may not be less expensive than pool storage for every licensee.

5.9.3 What is a dry-storage cask, and how does it keep the fuel from melting or from causing a nuclear reaction (criticality)?

Dry storage involves sealing used or spent fuel above ground in airtight steel (or in steel and concrete) containers or casks that provide both structural strength and shielding. The spent fuel is surrounded by inert gas inside the cask. All casks are passive designs in that they involve no mechanical devices for cooling or ventilation. Casks must be safety tested to withstand a variety of disasters, such as floods, projectiles originating from a tornado, temperature extremes, and lightning strikes. The cask is designed to preclude the possibility of an inadvertent criticality under all credible accidents or conditions. It must also provide adequate confinement, shielding, and heat removal during normal and accident conditions, including cask tipovers and drop accidents. Casks are placed vertically on a concrete pad or inside a concrete storage building or are inserted horizontally into a steel-reinforced concrete vault, depending on cask design. They will receive spent fuel that has been cooling in the reactor's spent fuel pool for at least 5 years. Depending on the design, casks hold from 7 to 68 spent fuel assemblies. The maximum amount of heat that is generated by the radioactive decay of fission products within the spent fuel cask will be less than that given off by 240 100-watt light bulbs, and this amount of heat will gradually decrease with time.

5.9.4 Who is responsible for reviewing proposed cask designs to ensure that they will safely confine the fuel, and what types of evaluations are required?

The NRC is responsible for reviewing proposed cask designs to ensure that they will safely confine the fuel and prevent fuel cladding (which surrounds the fuel) from degrading. The NRC regulations cover the testing, manufacture, and maintenance of casks used in dry storage. This includes an evaluation of natural events (earthquakes, high winds and tornadoes, wind-driven projectiles, flooding), an evaluation of accidents (explosions, fires, drops, tipovers, airplane accidents), and an evaluation of sabotage.

5.9.5 Are there any nuclear plants that already use dry-storage casks in an independent spent-fuel storage installation (ISFSI)?

The first dry-storage installation was licensed by the NRC in 1986. As of May, 2000, there were 12 nuclear power facilities using dry storage: Surry, Oconee, H.B. Robinson, Calvert Cliffs, Fort St. Vrain, Palisades, Point Beach, Prairie Island, Davis-Besse, Susquehanna, Arkansas Nuclear One and North Anna.

5.9.6 Can casks be used for both shipping and storage – or will they need to transfer the fuel from one cask to another at some later date?

Some casks are used solely for shipping spent nuclear fuel, some for storage purposes only, and some are dual purpose and can be used for both storage and shipment of the spent fuel. These casks are called dual purpose casks, and are licensed for both activities. If the spent fuel is stored in a cask that is not licensed for shipment, then it will need to be transferred to a cask licensed for shipment before it is removed from the site. Depending on the cask design, the transfer process can be done at the ISFSI site, or it may require a pool of water, such as the spent fuel pool.

5.9.7 How long may the licensee keep the spent fuel in an onsite ISFSI?

The NRC has determined that spent fuel can be stored onsite for at least 30 years beyond the licensed operating life of nuclear power plants—safely and with minimal environmental impact. This includes storage in the spent fuel pool or at either onsite or offsite independent spent fuel storage installations.

5.9.8 Can the spent fuel be shipped to another facility's ISFSI for storage?

Yes. Regulations allow for spent fuel from one facility to be stored in an ISFSI located at another facility with appropriate license conditions (this only applies if the second facility has a specific type of license called a "site specific Part 72" license in reference to the location in the regulations that describe the license requirements). The spent fuel would have to be transported in a cask that met NRC requirements for shipping casks.

5.9.9 Why is the licensing process on the independent spent-fuel storage installation (ISFSI) evaluated separately from the decommissioning process?

Both operating plants and plants that have permanently ceased operations and are decommissioning use ISFSIs. ISFSIs are not unique to decommissioning plants. The initial development of the decommissioning regulations occurred in the early 1980's. At that time the NRC and the industry assumed that by the time facilities begun decommissioning the U.S. Department of Energy's (DOE) high level and waste repository would be accepting spent fuel for ultimate disposal. Therefore, spent fuel onsite during decommissioning was not expected to be an issue. Consequently development of regulations related to ISFSI's occurred separately from the development of decommissioning regulations. Since the ISFSI may in some cases remain at the site longer than a nuclear facility that is undergoing immediate decommissioning, it is appropriate that ISFSI be capable of being licensed separately. The decommissioning of the ISFSI is also handled separately from the decommissioning of the nuclear power plant. Site specific ISFSI licenses requires the evaluation of the ISFSI separately from the remainder of the facility, although other site activities adjacent to the ISFSI are considered to evaluate their impact on the storage of the spent fuel. An ISFSI can be constructed and operated either under the same license that is used for the operating or decommissioning facility (called a "Part 50 license" in reference to the location in the *Code of Federal Regulations* that describes the license requirements), or under a site-separate license (called a "Part 72 license" in reference to the location in the *Code of Federal Regulations* that describes the licensing requirements for the ISFSI).

5.9.10 Where are the regulations relating to use of ISFSIs?

Regulations regarding the ISFSIs appear in the *Code of Federal Regulations*. The *Code of Federal Regulations* is a codification of the general and permanent rules published in the *Federal Register* by the executive departments and agencies of the Federal Government. The *Code* is divided into 50 titles, which represent broad areas subject to Federal regulation. Each title is divided into chapters; these usually bear the name of the issuing agency. Each chapter is further subdivided into parts covering specific regulatory areas. The regulations related to ISFSIs are in Title 10, "Energy," Chapter I – Nuclear Regulatory Commission, Part 72, "Licensing Requirements for the Independent Storage of Spent Nuclear Fuel and High-Level Radioactive Waste."

See the response to Question 13.6 for instructions on how to obtain a copy of the regulation.

5.9.11 Can the licensee use the site for other activities, such as a combustion turbine or natural gas plant, while fuel is stored onsite in an ISFSI?

The NRC will conduct a site-specific analysis if the licensee requests that the site be used for other activities, such as a combustion turbine or natural gas plant, while the fuel is stored onsite in an ISFSI. The decision as to whether to allow the alternate activities will depend on the results of the site-specific analysis.

6 RADIOACTIVE LOW-LEVEL WASTE

6.1 What is meant by low-level radioactive waste, and how is it different from fuel?

Low-level waste (LLW) is any radioactive waste that is not classified as high-level waste, spent nuclear fuel, transuranic waste (containing manmade elements heavier than uranium that emit alpha radiation—transuranic waste is produced during reactor fuel assembly, weapons fabrication, and chemical processing operations), or uranium or thorium mill tailings. LLW often contains small amounts of radioactivity dispersed in large amounts of material, but may also have activity levels requiring shielding and remote handling. It is generated by reactor fuel production, reactor operations, isotope production, medical procedures, and research and development activities. LLW usually comprises the following material contaminated with radionuclides: rags, papers, filters, solidified liquids, ion-exchange resins, tools, equipment, discarded protective clothing, dirt, construction rubble, concrete, or piping.

NRC regulations classify LLW on the basis of potential hazards, such as the concentrations of short-lived and long-lived radionuclides, in accordance with 10 CFR 61.55. Thus, LLW usually, but not necessarily, includes waste with relatively low concentrations of radionuclides. Although the classification of waste can be complex, Class A waste generally contains lower concentrations of longer half-life radioactive material than Class B and C wastes. Greater than Class C waste are not considered low-level radioactive waste and must be handled and disposed of differently from Class A, B and C wastes. See 10 CFR 61.55, "Waste Classification," for a detailed description of waste classification.

6.2 How is the low-level radioactive waste disposed of?

Low-level waste is commonly disposed of by burial in near-surface shallow trenches. After they are filled with containers, the trenches are usually covered with a low-permeability cover (such as clay). They are then often covered with a gravel drainage layer and a layer of topsoil. Vegetation is planted on top for erosion control. There is no intent to recover the wastes once they are disposed of. The volume of waste that is being disposed of each year is decreasing as the result of industry efforts to compact or incinerate part of the waste.

6.3 Is low-level waste (LLW) disposal safe?

LLW facilities are sited in areas that are away from surface water and where the groundwater is located at depths sufficiently beneath the trenches to minimize nuclide migration. Sites and the surrounding areas are monitored using a system of wells to determine if there is any leakage of radioactivity into the groundwater.

A combination of natural site characteristics and engineered safety features is used to ensure the safe disposal of LLW. In addition, restrictions of types and amounts of waste disposed of at a site, as well as the analysis performed as part of the licensing to demonstrate compliance with performance objectives in NRC regulations, increase the safety of LLW disposal.

The natural characteristics of an LLW disposal site are relied on in the long term, and they should promote disposal-site stability and attenuate the transport of radionuclides away from the disposal site into the general environment. Sites generally must possess the following characteristics: (1) relatively simple geology, (2) well-drained soils free from frequent ponding or flooding, (3) lack of susceptibility to surface geological processes, such as erosion, slumping, and landslides, (4) a water table of sufficient depth so that groundwater will not periodically intrude into the waste or discharge onsite, (5) lack of susceptibility to tectonic processes, (6) no known potentially exploitable natural resources, (7) limited future population growth or development, and (8) capability of not being adversely impacted by nearby facilities and activities.

Engineered barriers are manmade structures designed to improve the site's natural capability to isolate and contain waste. They consist of various engineered system components, including the following: (1) a layered earthen cover, (2) a disposal vault, (3) a drainage system, (4) waste forms and containers, (5) backfill material, and (6) an interior moisture barrier and low-permeability membrane.

Regulations specify the allowable radiation dose from the LLW facilities to the workers and to the public.

6.4 Where can low-level radioactive waste be disposed of?

There are currently three active, licensed disposal facilities. All three sites are located in Agreement States and are regulated by the States.[a] They are located in Barnwell (South Carolina), Hanford (Washington),

---

[a] The concept of Agreement states was set up by the Atomic Energy Act, and it permits NRC to relinquish to the States (on a state-by-state basis), certain of its authority to regulate specific areas,

31

and Clive (Utah). The site in Utah is restricted to specific types of low-level waste (LLW). The site in Washington State is restricted to waste from the Northwest and Rocky Mountain regional compacts. The site in South Carolina accepts LLW from all States except North Carolina. There are several additional sites currently under consideration for LLW disposal.

The Low-Level Radioactive Waste Policy Amendments Act of 1985 authorized the formation of regional compacts. A compact is a group of States (not necessarily with contiguous borders) that have decided to share resources to develop and maintain an low-level waste disposal site. Nine compacts are currently active, although most of them do not have a low-level waste site that has been developed or licensed. The Act contains a system of milestones, incentives, and penalties to ensure that States and compacts will be responsible for their own waste. Compacts can restrict access to the disposal site from States outside the compact.

6.5 What would happen if the waste site that was being used is closed?

If the low-level waste site is closed, the licensee would need to find another waste site that is willing to take the waste. Options such as compaction or incineration would be investigated as a means of reducing the magnitude of this waste stream. Until a new waste site is located, the licensee would need to temporarily store the waste onsite.

6.6 Can the low-level radioactive waste be stored at the site in the event that the waste site is closed? What type of facility is required and how long can the waste be left at the site?

The NRC has historically discouraged the use of onsite storage as a substitute for permanent disposal, but has not limited the amount of time that the waste can be stored. However, low-level waste (LLW) is normally stored onsite on an interim basis before being shipped offsite for permanent disposal. Onsite storage facilities are designed to minimize personnel exposure. High-dose-rate LLW is isolated in a shielded storage area and is easily retrievable. The lower dose-rate LLW is stacked or stored to maximize packing efficiencies. The NRC has guidelines regarding the storage facility, including the following: (1) shielding used should be controlled by dose rate criteria for both the site boundary and any adjacent offsite areas and (2) a liquid drainage collection and monitoring system should be present. The drain should be routed to a radwaste processing system.

6.7 Can radioactive waste be buried onsite?

Onsite burial is not acceptable unless (1) the licensee applies for permission and is specifically authorized under 10 CFR 20.2002 and (2) such burial is not prohibited by compact (that is, approved by the State).

6.8 Can the reactor vessel be disposed of at a low-level waste (LLW) site? Can the reactor vessel be shipped intact, or does it have to be segmented?

---

such as the disposal of low-level radioactive waste.

A reactor vessel can not be disposed of at a LLW site unless it meets waste classification requirements specified in the regulations and any site-specific requirements specified in the disposal facility's license. In most cases where disposal of the reactor vessel has occurred, the reactor vessel internals have been removed before any parts of the reactor vessel were shipped to an LLW disposal site. In two cases, the Trojan Nuclear Plant and the Saxton Facility, the reactor vessel was removed from the building and shipped intact to a LLW disposal site. Licensees are required to demonstrate that the shipment meets the regulations for package integrity and that the package meets the acceptance criteria for the LLW disposal site, such as criteria for radionuclide concentration and waste form.

6.9 How are liquid wastes disposed of?

Liquid wastes are processed onsite. The liquid portion is separated from the solid portion. The solid portion is disposed of in the low-level waste site, as long as it meets the criteria for low-level waste. The concentration of radioactive material in the liquid portion is measured, and if the concentration is below the Federal limits for release of effluents, the liquid portion may be released offsite (for instance, to sewers or a nearby body of water). Otherwise, the liquid portion is solidified (by mixing with concrete or similar solidifying or absorbing material) and disposed of as solid low-level waste.

6.10 Where are the regulations relating to radioactive low-level waste (LLW) disposal?

Regulations related to radioactive low-level waste disposal appears in the *Code of Federal Regulations*. The *Code of Federal Regulations* is a codification of the general and permanent rules published in the *Federal Register* by the executive departments and agencies of the Federal Government. The *Code* is divided into 50 titles, which represent broad areas subject to Federal regulation. Each title is divided into chapters; these usually bear the name of the issuing agency. Each chapter is further subdivided into parts covering specific regulatory areas.

The regulations related to LLW disposal are in 10 CFR Part 61 and 10 CFR Part 20 Subpart K. See the response to Question 13.6 for instructions on how to obtain a copy of the regulations.

7 TRANSPORTATION

7.1 How is the spent fuel going to be shipped to a final repository?

The spent fuel will be shipped in specially designed shipping casks. The design, construction, use, and maintenance of these commercial shipping containers are regulated by the NRC. In some cases, a dual-purpose cask, which is designed to work as a storage cask as well as a shipping cask, will be used. The shipping casks can be loaded onto trucks, trains, or barges for shipment.

7.2 How is low-level radioactive waste shipped to the disposal site?

Most low-level radioactive waste is shipped in packages authorized by the U.S. Department of Transportation, although some packages for larger quantities of low-level waste require NRC certification. Low-level radioactive waste packages can be loaded onto trucks or trains for shipment to the low-level waste site.

7.3 Are there regulations on radiation levels during transportation?

Yes. There are regulations governing the radiation level that stipulate limits (1) on the outer surface of the vehicle that is carrying the radioactive material, (2) at 6.6 feet (2 meters) from the surface of the vehicle, and (3) at the position occupied by the driver of the vehicle. Measurements of the applicable radiation levels are required before a vehicle is allowed to leave with packages.

7.4 Are there safety criteria for spent fuel shipping casks, and how are the criteria satisfied?

Yes. Safety standards for spent fuel shipping casks are detailed in NRC regulations. These regulations are compatible with safety standards issued by the International Atomic Energy Agency. Casks must be designed to withstand a series of tests that simulate accident as well as normal conditions of transportation. The normal conditions of transportation that must be considered include temperature, pressure, vibration, water spray, impact, penetration, and compression tests. The following tests are used to provide reasonable assurance that the casks will withstand serious transportation accidents:

- a 30-foot drop onto a flat, unyielding surface

- a 40-inch drop onto a vertical steel rod (puncture test)

- a 30-minute exposure to a fire of 1475°F

- submersion in 50 feet of water.

A cask design must be reviewed by the NRC staff to verify its resistance to accidents. Applicants must demonstrate to NRC that their design satisfies all applicable requirements. That demonstration may involve comparative evaluations with approved designs, analyses, and partial-scale tests. An approval certificate must be issued by the NRC before a particular cask design can be used to transport spent fuel.

7.5 Are there safety criteria for shipping containers for low-level waste (LLW), and how are the criteria satisfied before shipment?

Yes. LLW is shipped in containers that are designed to NRC or U.S. Department of Transportation (DOT) standards. Packaging requirements for shipments of specific radioactive materials are based on a number of factors, including the material activity, quantity, form (normal or special), specific activity, fissile properties, and other characteristics (physical, chemical, and nuclear properties). For smaller quantities of LLW, one of three types of containers is used, depending upon the material activity, or in some cases, the

34

specific activity. These container types are "strong tight containers," industrial packages, and Type A containers. Safety criteria for these types of containers, which are found in the DOT regulations, increase along with the package categories from "strong tight" to Type A containers. Type A containers must be able to withstand the normal conditions of transport as described in the answer to Question 7.4.

Wastes that contain higher levels of radioactivity are transported in Type B containers. Type B containers must be able to withstand the normal *and* accident conditions and are certified by the NRC as described in the response to Question 7.4.

For all types of containers, additional regulations apply that (1) specify preliminary determinations that must be ascertained before the first use of any container (for example, no cracks, no defects) and (2) specify routine determinations that must be satisfied before each shipment (for example, proper package for specific contents, properly installed closure devices).

7.6 What regulations apply to the transportation of radioactive material?

The transportation of radioactive materials is regulated jointly at the Federal level by the U.S. Department of Transportation (DOT) and the NRC. The responsibilities of the two agencies are delineated in a memorandum of understanding. (See *Federal Register* of July 2, 1979.) In general, the areas regulated by the agencies are as follows:

DOT – Regulates shippers and carriers of radioactive material and the conditions of transport (including routing, tiedowns, radiological controls, vehicle requirements, hazard communication, handling, storage, emergency response information, and employee training). DOT regulations are located in the *Code of Federal Regulations*, Title 49, "Transportation."

NRC – Regulates users of radioactive material and the design, construction, use, and maintenance of shipping containers used for larger quantities of radioactive material and fissile material (such as uranium). NRC regulations are located in the *Code of Federal Regulations*, Title 10, "Energy," Part 71, "Packaging and Transportation of Radioactive Material."

See the response to Question 13.6 for instructions on how to obtain a copy of the regulations.

7.7 Will the public be informed beforehand of low-level waste shipments or shipments of spent fuel and the routes taken?

No. The NRC does not monitor, or inform the public, or require the licensees to inform the public about the timing of shipments of low-level radioactive material wastes or the routes that the shipments will take. NRC does require licensees to provide advance notification to State governments regarding shipments of spent fuel and other specific types of waste shipments.

7.8 Are specific routes used for transporting radioactive material, and does the NRC approve the routes used for radioactive material shipments?

Specific routes are generally not required for transporting low-level radioactive materials. DOT's Federal Highway Administration has established routing requirements for spent fuel shipments. Essentially, these requirements limit these shipments to Interstate System Highways and city bypasses that minimize time in transit. Under NRC's regulations for physical protection of spent fuel, licensees are required to request and obtain advance approval from the NRC of the routes used for road and rail shipments of spent fuel and of any U.S. ports in which vessels carrying spent fuel shipments are scheduled to stop.

7.9 How safe are shipments of spent fuel casks? What would happen if the train or truck carrying a spent fuel cask was involved in an accident?

According to the analyses performed and the statistical record to date, shipments of spent fuel casks are very safe. There have been no recorded injuries or fatalities attributed to spent fuel in over 3,000 shipments in the U.S. over the last 30 years. In 1986, the NRC contracted with Lawrence Livermore National Laboratory to perform a study to determine the level of safety provided during shipments of spent fuel from U.S. commercial nuclear power plants. The study estimated that about one accident in every 80-million shipment miles could cause cask damage that would be significant enough to cause a radiological hazard that could equal or slightly exceed existing compliance values. Further information on this study can be found in NUREG/BR-0111, "Transporting Spent Fuel: Protection Provided Against Severe Highway and Railroad Accidents."

7.10 What emergency response plans are in place for transportation accidents involving a spent fuel cask or low-level-waste packages?

DOT requires carriers to receive emergency-response training that includes written procedures. In addition, DOT requires emergency-response procedures to accompany each shipment.

If an accident occurs, State and local governments are primarily responsible for overseeing the response of the carrier, shipper, and others, and for taking any actions deemed necessary to protect public health and safety.

The authorities that are likely to respond to transportation incidents (for example, police, fire fighters) have been provided with Emergency Response Guidebooks by DOT; these identify the potential hazards of materials and discuss corresponding mitigative actions. Each State has emergency plans in place for responding to incidents involving transportation of radioactive material. To assist the States in the development of these plans, a guide exists entitled "Guide and Example Plan for Development of State Emergency Response Plans and Systems for Transportation-Related Radiation Incidents"; it was prepared by the Western Interstate Nuclear Board and Regional Training Committee, Region VIII, Denver, Colorado, in 1975. An NRC- sponsored survey was conducted to ascertain the States' readiness to respond to radiological transportation incidents. In the responses to the survey, the States indicated that they have the necessary plans and procedures in place. This survey is documented in NUREG/CR-5399, "Survey of State and Tribal Emergency Response Capabilities for Radiological Transportation Incidents," which was prepared by Indiana University in Bloomington, Indiana, early in 1990.

To assist State and local governments, the Federal Government has established an interagency plan called the Federal Radiological Emergency Response Plan under the coordination of the U.S. Department of Energy (DOE), which charges eight regional coordinating offices with the responsibility and authority for convening radiological assistance teams. Upon request from a State, DOE will respond immediately to an emergency, including providing assistance at the scene.

8 LICENSE TERMINATION AND
THE ULTIMATE DISPOSITION OF THE FACILITY

8.1 How does decommissioning end, and who decides that the decommissioning is complete?

Licensees must submit an application for license termination at least 2 years before the requested termination date of the license. The license-termination plan must include

- a site characterization

- identification of remaining dismantlement activities

- plans for site remediation

- detailed plans for the final survey of residual contamination on the site

- a method for demonstrating compliance with the radiological criteria for license termination. For restricted release, the license-termination plan should include a description of the site's end use and documentation on public consultation, institutional controls, and financial assurance needed to comply with the requirements for license termination for restricted release or alternative criteria.

- an updated site-specific estimate of remaining decommissioning costs

- a supplement to the environmental report that describes any new information or significant environmental changes associated with the licensee's proposed termination activities.

After receiving the license-termination plan, the NRC places a notice of the receipt of the plan in the *Federal Register* and makes the plan available to the public for comment. The NRC also schedules a public meeting near the facility to discuss the plan's contents with the public. Because this is an action that involves a license amendment, there is also an opportunity for members of the public to request a hearing. If the license-termination plan demonstrates that the remainder of decommissioning activities will be performed in accordance with the NRC's regulations, is not detrimental to the health and safety of the public, and does not have a significant effect on the quality of the environment, then the Commission approves the plan by a license amendment, subject to whatever conditions and limitations the NRC deems appropriate and necessary. At this point, the licensee may achieve license termination by successfully demonstrating implementation of the license-termination plan, as approved by NRC.

The NRC will determine if the remaining dismantlement is performed in accordance with the approved license-termination plan. If this is the case, and if the final radiation survey and associated documentation demonstrate that the facility and site are suitable for release, then the Commission terminates the license, and the decommissioning process is considered to have been completed.

8.2 Why is the license-termination plan filed so late in the process?

The initial decommissioning activities (decontamination and dismantlement) are not significantly different from routine operational activities such as replacement or refurbishment. Because of the framework of regulatory provisions that are in place in the licensing basis for each facility, these activities do not present significant safety issues for which a detailed plan such as the license-termination plan is warranted. Therefore, it is appropriate that the licensee be permitted to conduct these activities without the need for a license amendment. At the license-termination stage (towards the end of the decommissioning process), the Commission must consider (1) the licensee's plan for assuring that adequate funds will be available for final site release, (2) the radiation-release criteria for license termination, and (3) the adequacy of the plans for the final survey that is required to verify that the release criteria have been met.

8.3 How will the licensee know where the radioactive material or contamination is located within the plant?

During operation, the plant is required to keep records of radiological surveillances that document where contaminations occur and locations of activation products and other sources of radiological materials. This is the basis for a plant's initial assessment of the location of radioactive material and contamination. Radiological surveys continue during all phases of decommissioning, and the records of these surveys are also kept. This information is used as part of the basis for the site-characterization plan; however, additional measurements are made at the site-characterization stage. The characterization plan must be designed to demonstrate compliance with appropriate dose or risk-based regulations.

8.4 What is included in the site characterization?

The purpose of the site characterization is to ensure that the final radiation surveys are conducted to cover all areas where contamination existed, remains, or has the potential to exist or remain as well as to provide data for planning further decommissioning activities. The site characterization contains a description of (1) the radiological contamination on the site before any cleanup activities associated with decommissioning took place, (2) a historical description of site operations, spills, and accidents, (3) a map of remaining contamination levels and contamination locations, and (4) a description of the survey instruments and supporting quality assurance practices used in the site-characterization program.

8.5 What does "suitable for release" mean? Are there any restrictions on how the site can be used?

There are two broad categories of uses for the facility after the license termination. The first is "unrestricted use," and the second is "restricted use." These will be discussed separately.

Unrestricted use means that there are no restrictions on how the site may be used. The licensee is free to continue to dismantle any remaining buildings or structures and to use the land or sell the land for any type of application.

Restricted use means that the licensee has demonstrated that further reductions in residual radioactivity would result in net public or environmental harm, or residual levels are as low as is reasonably achievable.

The licensee must have made provisions for legally enforceable institutional controls (for example, restrictions placed in the deed for the property describing what the land can and cannot be used for), which provide reasonable assurance that the radiological criteria set by the NRC will not be exceeded. In addition, the licensee must have provided sufficient financial assurance to an amenable independent third party to assume and carry out responsibilities for any necessary control and maintenance of the site. There are also regulations relating to the documentation of how the advice of individuals and institutions in the community who may be affected by the decommissioning has been sought and incorporated in the license-termination plan related to decommissioning by restricted use.

Although power reactor licensees can choose either a restricted or unrestricted option for release, the restricted option is primarily for materials licensees and would not normally be selected by reactor licensees because of the low levels of site contamination.

8.6 Why would the licensee be allowed to restrict use of the site?

There can be situations in which restricting site use can provide protection of public health and safety by reducing the total effective dose equivalent in a more reasonable and cost-effective manner than unrestricted site use. This protection is afforded by limiting access to the site, limiting the amount of time that an individual spends onsite, or by restricting agricultural or drinking water use. For many facilities, the time period requiring this type of restriction can be fairly short, and need only be long enough to allow radioactive decay to reduce radioactivity to levels that permit the site to be released for unrestricted use.

8.7 What is residual radioactivity, and why is it important to the termination of the license?

The term "residual radioactivity" means radioactivity in structures, materials, soils, groundwater, and other media at a site resulting from activities under the licensee's control. This includes radioactivity from all licensed and unlicensed sources used by the licensee, but excludes background (natural) radiation. It also includes radioactive materials remaining at the site as a result of routine or accidental releases of radioactive material at the site and previous disposals at the site. Criteria for the termination of the license are based on the residual radioactivity levels remaining at the site at the end of decommissioning.

8.8 What are the criteria for residual radioactivity at the site at the end of decommissioning, assuming that the licensee is planning for unrestricted use of the site?

The Commission has established a dose of 25 millirem (0.25 millisievert) per year total effective dose equivalent to an average member of the critical group as an acceptable criterion for release of any site for unrestricted use. The dose limit includes the dose from drinking groundwater. The licensee will be required to show that the site can meet this criterion before the license will be terminated for unrestricted use. In addition, the licensee will need to show that the amounts of residual radioactivity have been reduced to levels that are as low as reasonably achievable. This concept, known as ALARA, means that all doses are to be reduced below required levels to the lowest possible level considering economic and societal factors. Determination of levels that are ALARA must take into account consideration of any detriments, such as deaths from transportation accidents that are expected to potentially result from decontamination and waste disposal.

8.9 Why is there a difference between the residual radioactivity limits set by the NRC and those set by the EPA? Why do the licensees not have to use the limits set by EPA?

NRC continues to rely on the findings from two international organizations, the International Commission on Radiation Protection (ICRP) and the National Council on Radiation Protection and Measurements (NCRP). Both organizations have acknowledged the difficulty in setting acceptable levels of risk for the public; however, both ICRP and NCRP have established a dose of 100 mrem/yr to an individual member of the public as the level that is acceptable for exposure to radiation from sources other than medical procedures. The ICRP and the NCRP further established the need to reduce this annual dose rate by using the principle of "optimization," considering the cost effectiveness of additional dose reduction. Following these recommendations, the NRC adopted a level of 25 mrem/year as the value for residual radioactivity at a site under consideration for license termination.

EPA's radiation dose limit of 15 mrem/year results from a different technical analysis for establishing an acceptable risk to the public and a value for residual radioactivity other than that of NRC where radiation is the only contaminant considered. In addition, the NRC also has a "cleanup" requirement of "As Low As Reasonably Achievable" (ALARA) as discussed in the response to Question 8.8. The use of the ALARA requirement usually results in cleanups that are below the EPA's requirements also.

Nuclear reactors are licensed by the NRC, and the NRC is responsible for making the safety and environmental determination for termination of the license. Therefore, licensees are required to meet the NRC's requirements for residual radioactivity. However, since the NRC value of 25mrem/year is a limit a licensee can choose to further reduce the value of residual radioactivity at a site to achieve annual dose values less than 25 mrem/year.

8.10 What is a "total effective dose equivalent?"

The total effective dose equivalent is a term that is used to express how the radiation dose is calculated to an individual. It means that the dose from radioactive material outside of the individual (external radiation) and the dose from any radioactive material that the individual may have inhaled or ingested (internal radiation) have been considered. For the latter case, the internal radiation dose is considered for a period of 50 years following the intake of the radioactive material. In addition, weighting factors are used that are specific to the body organs or tissues that are irradiated. These weighting factors are used to account for the variation in sensitivity of different organs or tissues to radiation.

8.11 Who would be considered an "average member of the critical group?"

The "critical group" means the group of individuals reasonably expected to receive the highest exposure to residual radioactivity within the assumptions of a particular scenario. The average dose to a member of the critical group is represented by the average of the doses for all members of the critical group, which in turn is assumed to represent the most likely exposure situation. For example, when considering whether it is appropriate to "release" a building (allow people to work in the building without restrictions) that has been decontaminated, the critical group would be the group of regular employees that would work in the building. If radiation in the soil is the concern, then the scenario used to represent the maximally exposed individual is that of a resident farmer. The assumptions used for this scenario are "prudently conservative" and tend to overestimate the potential doses. The added sensitivity of certain members of the population, such as pregnant women, infants, and children, are accounted for in the analysis. However, the most sensitive member may not always be the member of the population that receives the highest dose. This is especially true if the most sensitive member (for example, an infant) does not participate in specific activities that may provide the greatest dose or if they do not eat specific foods that cause the greatest dose.

8.12 Have any facilities been decontaminated to the point of being allowed unrestricted release?

Yes, two commercial reactors, Shoreham and Fort St. Vrain, have had their licenses terminated and the sites released for unrestricted use. The Pathfinder reactor has had its license terminated, and most of the site has been released for unrestricted use. In addition, many non-reactor facilities, such as hospitals and industrial and pharmaceutical manufacturers have, had their licenses terminated and the site returned to unrestricted use.

8.13 What are the criteria for residual radioactivity at the site at the end of decommissioning, assuming that the licensee is planning for restricted use of the site?

The Commission has also established criteria for restricted use of the site. These are more complex than for unrestricted use and employ a tiered approach. Residual radioactivity at the site must have been reduced so that there would be reasonable assurance that the total effective dose equivalent from the residual radioactivity to the average member of the critical group would not exceed 25 millirem (0.25 millisievert) per year with institutional controls in place and either 100 millirem (1 millisievert) per year or 500 millirem (5 millisievert) per year with no institutional controls. Institutional controls include such engineered controls as fences and such restrictions on the site's deed that activities like a park or farming would not be allowed. Institutional control could also include ownership by the Federal or State government, thus providing for a legal mechanism to restrict public access.

For the first case, 100 millirem (1 millisievert) per year, the licensee must demonstrate the following:

- Further reductions in residual radioactivity would result in net public or environmental harm, or they were not made because the residual levels are as low as reasonably achievable, taking into account the consideration of any detriments, such as traffic accidents, that may be expected to potentially result from decontamination and waste disposal.

- Provisions have been made for legally enforceable institutional controls to provide assurance that the 25 millirem (0.25 millisievert) per year average dose to the average member of the critical group will not be exceeded.

- Funds have been placed into an account segregated from other assets and outside of the licensee's administrative controls that will be used to pay for any necessary control and maintenance of the site, or that a surety method, insurance, or other guarantee method has been established.

- The licensee has sought advice from affected parties and in seeking that advice provided for (1) participation by representatives of a broad cross section of community interests, (2) an opportunity for a comprehensive collective discussion on the issues, and (3) publicly available summary of the results of all such discussions.

For the second case of 500 millirem (5 millisievert) per year, the licensee must demonstrate the following:

- Further reductions in residual radioactivity necessary to comply with the 100-millirem (1-millisievert)-per-year value are not technically achievable, are prohibitively expensive, or would result in net public or environmental harm.

- Provisions have been made for legally enforceable and durable institutional controls (which may also include Federal, State, or local government control of sites), as well as provisions for a verification of the continued effectiveness of the institutional controls at the site every 5 years after license termination to ensure that the institutional controls are in place, and the restrictions are working.

- Sufficient financial assurance has been provided to enable a responsible government entity or independent third party to carry out periodic rechecks of the site no less frequently than every 5 years to ensure that the institutional controls remain in place as necessary to meet the 25-millirem (0.25-millisievert)-per-year criterion. Sufficient financial assurance must also be provided to assume and carry out responsibilities for any necessary control and maintenance of those controls.

- The licensee has sought advice from affected parties and in seeking that advice provided for (1) participation by representatives of a broad cross section of community interests, (2) an opportunity for a comprehensive collective discussion on the issues, and (3) a publicly available summary of the results of all such discussions.

In addition, alternate criteria exist for the case in which the 25-millirem-per-year limit is found to be inappropriate. In this situation, it must be unlikely that the dose from all manmade sources combined, other than medical, would be more than 100 millirem (1 millisievert) per year. These alternate criteria are expected to be used only in rare cases. The licensee must

- submit an analysis of possible sources of exposure to provide assurance that public health and safety would continue to be protected

- demonstrate that it has employed, to the extent practical, restrictions on the site use

- reduce doses as low as reasonably achievable, taking into consideration any detriments, such as traffic accidents, that are expected to potentially result from decontamination and waste disposal

- submit a license-termination plan, which specifies that it plans to decommission by using alternate criteria and documents how the licensee has sought and addressed advice from affected parties and in seeking that advice provided for (1) participation by representatives of a broad cross section of community interests, (2) an opportunity for a comprehensive collective discussion on the issues, and (3) a publicly available summary of the results of all such discussions.

The use of alternate criteria to terminate a license requires the approval of the Commission after a consideration of the NRC staff's recommendations that address any comments provided by the U.S. Environmental Protection Agency or the public.

8.14 How does the dose based on the residual radioactivity levels relate to background dose levels?

This dose can be compared with the background dose of approximately 300 millirem (3 sievert) per year that is anticipated to the average person living in the United States. Background radiation means radiation from cosmic sources, naturally occurring radioactive material, including radon, and global fallout as it exists in the environment from the testing of nuclear explosive devices or from earlier nuclear accidents, such as Chornobyl, that contributes to background radiation and is not under the control of the licensee. "Distinguishable from background" means that the detectable concentration of a radionuclide is statistically different from the background concentration of that radionuclide in the vicinity of the site.

8.15 Why didn't the NRC set the final dose criteria for release of the site for unrestricted use to "pre-existing background" levels?

For those facilities in which soil or building contamination exists, it would be extremely difficult to demonstrate that an objective of "return to background" had been achieved. In addition, the removal of soil or concrete to "pre-existing background" levels is generally not desirable from the perspective of risk to public health and safety and protection of the environment. For example, at some point, the removal of increasingly larger volumes of concrete and soil would also result in a greater net risk from transportation accidents.

8.16 Is it possible that some isotopes are located in such a way that radiation-monitoring devices cannot accurately detect their levels of radioactivity?

It is unlikely that radioactive material located inside a piece of equipment or a structure is not detected during the final radiation survey. The structures, systems, and components that have radioactive contamination exceeding NRC's limits will be decontaminated or dismantled and shipped to a low-level-waste disposal site. The licensee must keep records of information during the operating phase of the facility that could be used to identify where spills or other occurrences involving the spread of contamination in and around the facility, equipment, or site have been located.

8.17 Will continued monitoring be required after the decommissioning process is complete to ensure that the radiation levels do not increase?

No. For sites that have been determined to be acceptable for unrestricted use, there are no requirements for further measurement of radiation levels. It is not expected that these radiation levels would change—other than to be reduced over time—because the radioactive material will have been removed from the site, and there would be no mechanism for further contamination or radiological releases.

For sites that have been determined to be acceptable for license termination under restricted conditions, additional measurements of radiation are only required for sites that have residual radioactivity in excess of 100 millirem (1 millisievert) per year, but less than 500 millirem (5 millisievert) per year. These measurements are to be made by a responsible government entity or independent third party, including a governmental custodian of a site. The measurements are to be carried out no less frequently than every 5 years to ensure that the institutional controls remain in place as necessary to meet the criterion of 25 millirem (0.25 millisievert) per year to an average member of the critical group.

8.18 What types of uses can be made of the plant site after decommissioning is completed?

Once the license has been terminated and the site released for unrestricted use, there are no restrictions on the type of use. Possible uses could range from restoring the natural habitat, to farming, to continued use as an industrial site (possibly leaving buildings and installing a gas-, coal-, or oil-powered generating plant). If a site is being decommissioned under restricted-release or alternate-release criteria, then NRC approval must be obtained, on a case-by-case basis, for the future uses of the plant site before the license is terminated.

8.19 Could the licensee initiate an alternative use of the site or partial site release before the decommissioning process is completed?

Requests by licensees to initiate alternative uses of the site or site partial release before the decommissioning process is completed would be reviewed by the Commission on a case-by-case basis. The NRC staff has proposed a rulemaking plan that would procedurize the process by which a licensee would release part of its reactor facility or site for unrestricted use before receiving review and approval of its license termination plan.

8.20 What uses have been made of sites that were decommissioned in the past?

The licensee that held the license for the Fort St. Vrain nuclear plant in Colorado has chosen to build a natural-gas-powered boiler for use with the existing turbine generator. The Pathfinder site has a natural-gas electric-generating plant. The Shoreham site is currently not being used.

8.21 What regulations are related to license termination?

Regulations regarding license termination appear in the *Code of Federal Regulations*. The *Code of Federal Regulations* is a codification of the general and permanent rules published in the *Federal Register* by the executive departments and agencies of the Federal Government. The *Code* is divided into 50 titles, which represent broad areas subject to Federal regulation. Each title is divided into chapters; these usually bear the name of the issuing agency. Each chapter is further subdivided into parts covering specific regulatory areas.

The regulations related to license termination are in Title 10, "Energy," Part 20, "Standards for Protection Against Radiation," Subpart E, "Radiological Criteria for License Termination." See the response to Question 13.6 for instructions on how to obtain a copy of the regulations.

9 HAZARDS ASSOCIATED WITH DECOMMISSIONING

9.1 Workers

9.1.1 Where do the decommissioning workers come from?

The majority of workers for an immediate decontamination and dismantlement program will likely be people who worked on the operating plant. These workers are most familiar with the facility and its history. Some jobs, however, may be contracted out to companies that have gained experience at other plants in specialized areas of decommissioning or dismantlement. There will be very few employees during the storage phase in facilities that are placed in a storage mode. A new group of workers will likely need to be hired who are most likely unfamiliar with the plant, but who will probably have had some decommissioning experience at other facilities.

9.1.2 Is worker safety considered in the planning for and review of decommissioning?

Yes. Worker safety is considered both in terms of the radiological hazard (their exposure to radiation) and in terms of industrial safety.

9.1.3 How much occupational dose is received by workers during decommissioning?

The amount of occupational dose received during the decommissioning process will depend on the design and size of the facility as well as on the plans for decommissioning. A greater amount of occupational dose is anticipated to be incurred for an immediate decontamination and dismantlement than for a storage period followed by dismantlement. Estimates were given in NUREG-0586, "Final Generic Environmental Impact Statement on Decommissioning of Nuclear Facilities," (NRC 1988) that ranged from 333 person-rem for a 30-year storage period to 1874 person-rem for an immediate decontamination and dismantlement. This can be compared with the 1996 annual average for an operating plant: 126 person-rem for pressurized-water reactors and 235 person-rem for boiling-water reactors. The person-rem numbers are the doses that are received by all the workers. The dose to any one worker is expected to be below the 5-rem-a-year regulatory limit and is usually well below this limit.

Since that study was performed, estimates for occupational dose from decommissioning range from 591 person-rem for the Trojan nuclear plant to 1215 person-rem for Maine Yankee (includes the dose from transportation of the low-level waste [LLW] and 996 person-rem for Haddam Neck (includes the occupational dose from the transportation of the LLW). All three of these plants are using an immediate decontamination and dismantlement type of decommissioning.

9.1.4 Are there limits on the amount of occupational dose that may be received?

45

Yes. The regulations state that the licensee shall control the occupational dose to individual adults to an annual limit of 5 rem (total effective dose equivalent to the entire body) or to an organ dose equal to 50 rem. There are also dose limits to the eyes, the skin, and the extremities.

9.1.5 Does the licensee have to estimate the occupational dose before the decommissioning process is initiated?

No. However, at the time that the license-termination plan is updated, the licensee is required to update its environmental report as appropriate to reflect any new information or significant environmental change associated with the applicant's proposed decommissioning activities. The environmental report contains an estimate of occupational dose, so the licensee needs to estimate the occupational dose for decommissioning to determine if the estimates are within the range given in the environmental report for routine operations.

9.2 Public and Environment

9.2.1 Is the safety of the public considered in the planning for and review of decommissioning?

Yes. The safety of the public is a major concern, even though the potential for hazards to the public from the decommissioning process and potential accidents is much less than it is when the facility is operating.

9.2.2 How much dose will the public receive during the decommissioning process?

The major source of exposure to the public is from the shipment of low-level waste from the reactor site to the low-level waste disposal site. Estimates made in a generic study of nuclear power reactor decommissioning (the GEIS, NUREG-0586) range from 3 person-rem for a 30-year storage period to 21 person-rem for immediate decontamination and decommissioning. The estimated public dose from the Trojan nuclear plant decommissioning is 4.8 person-rem. The estimated dose to the public from decommissioning the Haddam Neck plant is 11 person-rem. The radiation dose is received by people who travel along the same route as the trucks that are transporting low-level waste. However, because of the variability in the timing of each shipment, the short period of time that any person would be near any of the trucks, and the small dose that is allowed 6 feet from the side of a truck (10 millirem per hour), the dose that is received by any one person traveling down the highway or stopped at a rest stop is a very small fraction of the annual dose that the person would receive from background radiation.

Minor sources of exposure to the public include radioactive effluent releases during the decommissioning process as discussed in the response to Question 9.2.4.

9.2.3 Who estimates what the doses are, and how are these estimates made?

The licensee estimates the doses. The doses are estimated using assumptions about the amount of radioactive material that will be released or the proximity of the public to the source of radiation. The doses are calculated using NRC-approved scenarios, assumptions, parameter values, and conceptual models. The NRC reviews the licensee's estimates of the doses and often recalculates the doses using its own assumptions for activities with the potential for significant worker exposure.

9.2.4 What types of effluent releases are expected, and where will they enter the environment?

Three important radiation-exposure pathways need to be considered in the evaluation of the radiation safety of normal reactor decommissioning operations: inhalation, ingestion, and external exposure to radioactive materials. During decommissioning, inhalation is considered to be the dominant pathway of public radiation exposure, since exposure to radioactive surfaces and ingestion can be minimized or eliminated as radiation pathways to the public. During the transport of radioactive wastes, inhalation and ingestion can be minimized or eliminated as radiation pathways to the public by containing the waste in a form or a container that does not allow for release to the air or water. Therefore, for transportation of radioactive waste, external exposure to radioactive materials is considered to be the only pathway of concern.

During decontamination and dismantlement activities (DECON), radioactivity in air effluents is expected to be limited primarily to airborne radioactive particulates. The particulates will be filtered through high-efficiency particulate air (HEPA) filters in the ventilation systems of containment buildings, auxiliary buildings, and fuel-handling buildings. These are the buildings that contain the major sources of radioactive materials at a nuclear power facility. Air effluents from these buildings are monitored because the NRC has set limits on the amount of radioactive material that may be released. During storage (SAFSTOR), there should be little or no effluent releases since there are no decommissioning activities being performed. However, monitoring still continues to verify that any releases are minor.

Liquid radioactive wastes are collected, stored, and processed, depending on the amount of particulates and the source of the liquid wastes. The liquid waste is processed before any releases are made to the environment. There are limits on the amount of radioactive material or other types of wastes that may be released to the environment. The regulatory limits for radioactive discharges can be found in 10 CFR Part 20. Additionally, there are limits on the discharge of non-productive wastes. These limits are described on the National Pollutant Discharge Elimination System waste discharge permit that is issued to the specific facility. These permits are regulated by the States or the U.S. Environmental Protection Agency.

Direct exposure can occur during both SAFSTOR and DECON. Public exposure would only result if members of the public were close enough to the source of radiation to receive a dose. The farther away a person is from radioactive material, the smaller the dose. The site boundary determines how far from the source people should be to avoid direct exposure; at that distance, exposure would be negligible. However, the dose from a shipment of nuclear waste could be 10 millirem per hour at a distance of 6 feet from the side of the truck. This means that a person standing there for 1 hour would receive a maximum dose of 10 millirem, which is a very small fraction of the average annual exposure to background radiation.

9.2.5 Can you measure the effluent release to know how much is really entering the environment?

Yes. Air effluent releases and liquid releases from all licensed light-water power reactor sites are monitored in accordance with the licensee's Offsite Dose Calculation Manual (ODCM). Release limits given in the ODCM are based on NRC regulations. Special filters are used to collect particulate materials being released from the plant. Special air samplers and monitors are used to determine the amount of material that is released to the environment. Similar monitors are used to determine the amount of material released in liquid effluents.

9.2.6 Will there be continued environmental monitoring of the site and the offsite areas to measure releases of radioactive material during the decommissioning process?

Yes. The radiological environmental monitoring program that was in place at the nuclear plant will continue even after the plant is shut down. The program will be modified to appropriately monitor the types of releases that may occur during decommissioning and to monitor results at appropriate intervals of time. Not all measurements will be made on a continuous basis. The licensee uses the results of the environmental monitoring program to calculate the dose to the public. They follow a procedure that they have developed that is in a document called the Offsite Dose Calculation Manual.

9.2.7 Who will perform the environmental monitoring?

The Radiological Environmental Monitoring Program is conducted by the licensee. The procedures and results of the Radiological Environmental Monitoring Program are inspected and reviewed by NRC staff to ensure that all requirements are being met. The NRC does not independently monitor radioactive releases on a continuous basis. However, the NRC does occasionally have an independent measurement of the releases to verify that the licensee's program is adequate to analyze samples and to measure the amount of radioactive material that leaves the plant.

9.2.8 Are the monitoring reports available to the public? How would we find them?

Yes, each site prepares an annual environmental monitoring report that describes the Radiological Environmental Monitoring Program and the results of the program for that calendar year. The reports are available to the public through the Public Document Room and the Public Electronic Reading Room. See the responses to Questions 13.1 and 13.2 for instructions to access these sources of information.

9.2.9 What types of chemicals will be released to the environment?

Chemicals that are used at the operating facility and chemicals that are used specifically during the decontamination process must be disposed of in accordance with appropriate Federal and State regulations. Liquid releases from the plant to rivers, streams, or lakes must meet the requirements given in the facility's National Pollutant Discharge Elimination System permit, which is issued by the State or by the U.S. Environmental Protection Agency. Release of chemicals to waste sites must meet the requirements of that waste site. Mixed wastes (hazardous chemicals and radioactive material) are generally shipped to a licensed mixed waste processor.

9.2.10 What hazards are presented to the public when the waste is shipped?

Very minor amounts of radiation will be received by people in other vehicles driving alongside, or parked beside, trucks carrying low-level radioactive waste from the facility to the low-level waste site. The hazard from transportation accidents is discussed in responses to Questions 7.5 and 7.9. However, there is also a potential hazard resulting from the release of radioactive material due to a major accident involving a vehicle carrying radioactive waste. The low-level waste is shipped in a solid form, so any danger resulting from a typical vehicular accident would likely be confined to a small radius around the accident.

9.2.11 What types of accidents at the reactor site are considered and what would be the consequences to the public?

Once the reactor permanently shuts down, the risk to the public is greatly reduced; however, there are still several accidents that may occur with consequences offsite. The accidents that have the potential for the greatest offsite doses are those that involve the spent fuel that has recently been moved from the reactor to the spent fuel pool. Over time, the hazard from the spent fuel diminishes as the radioactive material in the fuel decays.

Licensees are required to examine their sites and decommissioning plans to identify postulated accidents that could occur during decommissioning. An analysis of these accidents is required in their Final Safety Analysis Report, which is part of the licensing basis for the plant. Except for the fuel-related accidents in the first year(s) after the facility ceases operation, the offsite consequences of these accidents are very small and do not require offsite emergency response. Examples of the types of accidents that are considered by the licensees include

- Cask or heavy load-handling accident with a subsequent drop into spent fuel pool

- Loss of cooling for the spent fuel pool or loss of water from the spent fuel pool

- Materials handling event (non-fuel)

- Radioactive liquid waste releases

- Accidents from handling spent resin

- Fire

- Explosions

- External events

- Transportation accidents.

If a licensee requests an exemption to a regulation because they believe it no longer applies due to the decommissioning state of the plant, they must show that the regulation is not needed including consideration of the risk to the public. Additional information regarding the consequences of spent fuel pool accidents is given in the response to Question 5.8.4.

9.3 General

9.3.1 In general, how safe is a decommissioning plant in contrast to an operating plant?

At the time that the plant permanently ceases operations and the fuel is removed from the reactor, the risk to the public from an accident drops significantly.

9.3.2 Will there still be emergency preparedness plans and warning sirens in the vicinity of the plant?

For some period of time after the licensee ceases reactor operations, the offsite emergency preparedness planning will be maintained. This period of time depends on when the reactor was last critical as well as onsite-specific considerations. Offsite emergency planning may be eliminated when the fuel has been removed from the reactor and placed in the spent fuel pool, and sufficient time has elapsed, and there are no longer any postulated accidents that would result in offsite dose consequences that are large enough to require offsite emergency planning. There would be no requirement to maintain offsite systems to warn the public. Onsite emergency plans will be required for both the spent fuel pool and the Independent Spent Fuel Storage Installations, but offsite plans will not be required.

If, however, an operating plant is located at the same site as the decommissioning plant, the emergency preparedness plans will still be in effect for the operating plant.

9.3.3 What measures are taken to prevent vandalism and sabotage during decommissioning?

The facility is required to have a security plan when a plant is being decommissioned; however, as the hazards are removed from the nuclear reactor site, security requirements are modified. Many plants reduce the area that they keep very secured to the area around where the spent fuel is stored. This is known as the "nuclear island." Reducing the size of the area that has strict security measures allows for better control of the material that must be safeguarded. Security measures for the nuclear island are designed to prevent sabotage or removal of the nuclear material.

10 FINANCES

10.1 How much does it cost to decommission a nuclear power plant?

The total cost of decommissioning depends on many factors, including the sequence and timing of the various stages of the program, location of the facility, current radioactive waste burial costs, and plans for spent fuel storage. The minimum amounts that are required for reasonable assurance of funds for decommissioning are $290 million for pressurized-water reactors and $370 million for boiling-water reactors. These costs are in 1999 dollars and are adjusted annually, as further specified in the regulations. These are minimum amounts to show reasonable assurance, rather than estimates, of what it would cost to decommission a specific nuclear reactor.

Actual site-specific costs incurred and estimated costs of decommissioning give a better indication of what the process costs. The Fort St. Vrain nuclear plant, which was a 330-megawatt-electric high-temperature gas-cooled reactor, ceased power operations in 1989 and underwent immediate decontamination and dismantlement. The decommissioning effort was completed in late 1996, and the license was terminated. The total cost of decommissioning was $189 million.

The cost for decommissioning the Trojan nuclear plant (an 1130-megawatt-electric pressurized-water reactor) is estimated to be on the order of $210 million in 1993 dollars, which does not include $42 million for non-radioactive site remediation or $110 million for the independent spent fuel storage installation and related fuel management. The Trojan nuclear plant is also planning an immediate decontamination and decommissioning from shutdown in 1993 to license termination in 2002.

The estimated cost for decommissioning the Haddam Neck nuclear plant, a 619-megawatt- electric pressurized-water reactor, is $344.4 million in 1996 dollars, not including $82.3 million in spent fuel storage costs (for a total of $426.7 million).

The estimated cost for decommissioning Maine Yankee, an 830-megawatt-electric pressurized- water reactor, is $274.9 million in 1997 dollars. This does not include costs for spent fuel management ($53.4 million) or for site restoration ($49.2 million), for a total of $377.6 million.

The estimated cost for decommissioning Big Rock Point, a 67-megawatt-electric boiling-water reactor, is $290 million in 1997 dollars.

The estimated cost for decommissioning Rancho Seco, a 913-megawatt-electric pressurized-water reactor is $441 million in 1995 dollars.

The estimated cost for decommissioning Yankee Rowe, a 175-megawatt-electric pressurized- water reactor is $306.4 million in 1995 dollars.

Decommissioning costs vary, based on plant size and design, local labor and radiological waste burial costs, and the specific process that is being used for decommissioning.

10.2 Who makes the estimates of the decommissioning costs?

The licensee makes the site specific estimates of the decommissioning costs or hires a contractor who has extensive experience in making these estimates. The site specific estimates are reviewed by the NRC.

10.3 When are the estimates of the decommissioning costs made?

The NRC has regulations regarding the methods used to reasonably ensure that funds will be available to decommission the facility. The NRC has specified a table of minimum amounts required to demonstrate reasonable assurance of funds for decommissioning by reactor type and power level ($105 million for pressurized-water reactors and $135 million for boiling-water reactors in 1986 dollars). The licensees must adjust this rate annually. Licensees may also perform site-specific estimates that could result in cost estimates that are higher than the generic-formula amounts specified in 10 CFR 50.75 (c).

An estimate is made at or about 5 years preceding the projected end of operations. At this time, power reactor licensees shall submit a preliminary decommissioning cost estimate, which includes an up-to-date assessment of the major factors that could affect the cost to decommission. If the amount of money available is inadequate, the licensee has approximately 5 years to adjust the money in the decommissioning trust fund to ensure that appropriate funds are available for decommissioning.

An estimate is submitted at the time that the post-shutdown decommissioning activities report (PSDAR) is submitted (no later than 2 years following permanent cessation of operations). This estimate may be (1) a site-specific cost estimate that is based on the activities and schedule that are also discussed in the PSDAR, (2) an estimate based on actual costs at similar facilities that have undergone similar decommissioning activities, or (3) a generic cost estimate. The NRC recommends that licensees planning an immediate decontamination and dismantlement submit a site-specific cost estimate in the PSDAR; however, a more generic one would be acceptable for facilities that are submitting their PSDAR in advance of the 2-year requirement. If a storage period is planned during decommissioning, the licensee should provide a method of adjusting the cost estimate and funding throughout the duration of the storage.

The regulations also require a site-specific cost estimate within 2 years following permanent cessation of operations if one has not already been submitted.

Finally, at the time that the license-termination plan is submitted (at least 2 years before the date when the license terminates), an updated site-specific estimate of any remaining decommissioning costs is required.

10.4 If the first estimate of decommissioning costs is made at the time that the facility is licensed, are there methods for adjusting for inflation?

NRC regulations provide an adjustment factor for cost escalation that takes into account escalation factors for labor, energy, and waste burial. The labor and energy escalation factors are obtained from regional data issued by the U.S. Department of Labor's Bureau of Labor Statistics. The waste-burial cost-escalation factor is taken from an NRC report, "Report on Waste Burial Charges" (NUREG-1307).

10.5 How does the NRC ensure that the licensee will have the money when it is needed for decommissioning?

Financial assurance is provided by the following methods:

Prepayment. In this case, at the start of operations, the licensee deposits into an account enough funds to pay the decommissioning costs. The account is segregated from the licensee's other assets and remains outside the licensee's administrative control of cash or liquid assets. Prepayment may be in the form of a trust, escrow account, government fund, certificate of deposit, or deposit of government securities.

External sinking fund. An external sinking fund is a fund established and maintained by setting funds aside periodically into an account segregated from licensee assets and outside the licensee's administrative control. The total amount of these funds would be sufficient to pay decommissioning costs at the time that it is anticipated that the licensee will cease operations. An external sinking fund may be in the form of a trust, escrow account, government fund, certificate of deposit, or deposit of government securities.

Surety method, insurance, or other guarantee method. A surety method may be in the form of a surety bond, letter of credit, or line of credit. Any surety method or insurance used to provide financial assurance must be open-ended, or if written for a specific term, such as 5 years, must be renewed automatically. An exception is allowed when the issuer notifies the Commission, the beneficiary, and the licensee of its intent to not renew within 90 days or more preceding the renewal date. The surety or insurance must also provide that the full face amount be paid to the beneficiary automatically preceding the expiration date without proof of forfeiture if the licensee fails to provide a replacement acceptable to the Commission within 30 days after receipt of notification of cancellation. In addition, the surety or insurance must be payable to a trust established for decommissioning costs, and the trustee and trust must be acceptable to the Commission. The surety method or insurance must remain in effect until the Commission has terminated the license.

10.6 Does the NRC have a method to monitor the decommissioning trust fund moneys for each specific nuclear power facility?

The NRC monitors the decommissioning funds. The regulations regarding how the funds are to be accumulated are very specific. Accounts for decommissioning are segregated from the licensee's other assets and remain outside of the licensee's administrative controls as discussed in the response to Question 10.5 (although there is a provision for combining the costs for long-term storage of the spent fuel onsite with the costs for decontamination and dismantlement of the facility as long as the funds for each are clearly identified). In addition, the regulations require that the licensee report at least once every 2 years on the status of its decommissioning funding for each reactor or part of a reactor that it owns. This same report is required annually for facilities that have already permanently ceased operations, or facilities that are within 5 years of the projected end of their operation, or if it appears that they might be within 5 years of shutting down.

10.7 Are the financial assurance reports available to the public?

The financial assurance reports that are discussed in the previous section and that are submitted to the NRC on an annual or biannual basis are part of the public record and can be obtained from the Public Document Room or the Public Electronic Reading Room by the process described in the response to Question 13.2 or 13.4.

10.8 Do the financial assurance regulations apply for Federal government licensees?

Certain Federal government licensees may provide a statement of intent containing a cost estimate for decommissioning and indicating that funds for decommissioning will be obtained when necessary as long as the obligations of such licensees are granted by and supported by the full faith and credit of the U.S. government.

10.9 Is there any way to ensure that the licensee does not just spend all of the money in the first few years of decommissioning and have nothing left to complete the job?

The NRC has placed regulations regarding the amount of money that can be used from the decommissioning fund at various stages of the decommissioning process. The licensee is allowed to use 3 percent of the generic amount of funds that are specified in the regulations for power plants based on their size and type for decommissioning planning that may occur, even while the facility is still operating. Appropriate activities include engineering design, work package preparation, and licensing activities.

After submitting the certification of permanent cessation of operations and the certification that the fuel has been removed from the reactor vessel, the licensee may use an additional 20 percent of the funds for any legitimate decommissioning activities. The licensee is prohibited from using the remaining 77 percent of the generic decommissioning funds until a site-specific cost estimate is submitted to the NRC. This cost estimate must be submitted within 2 years following permanent cessation of operations. Also, the NRC's regulations require licensees to use funds from decommissioning trusts only for those legitimate decommissioning activities consistent with the NRC's definition of decommissioning (see the response to Question 1.1).

10.10 What would happen if the cost of decommissioning exceeds the amount of money in the trust fund?

The various cost estimates (at the time of licensing, 5 years before anticipated shutdown, with the Post-Shutdown Decommissioning Activities Report submittal, 2 years following shutdown, and 2 years preceding the anticipated termination of the license) are a method of re-evaluating the decommissioning costs at various times and stages in the facility's life to ensure that there will be adequate funds available to complete the decommissioning process. If there is insufficient money in the trust fund, licensees would be required to obtain the additional funds needed to complete the decommissioning of the facility.

10.11 What would happen if the plant had an accident, and there was not enough money in the decommissioning trust fund to complete decommissioning and cleanup after the accident?

Licensees are required to carry insurance, which is separate from the decommissioning funding requirements, in an amount that would allow cleanup of the site to such a level that decommissioning could be completed with the full amount of the decommissioning trust fund. Currently, $1.06 billion per operating unit is required for such insurance coverage.

10.12 Will enough money still be in the decommissioning trust fund even if the facility ceases operation?

In the event that a facility shuts down prematurely before the decommissioning trust is fully funded, the licensee is still required to fully fund the decommissioning. Most utilities are diversified and are able to recapture these costs as part of their continued business. To date, none of the utilities of the prematurely shut-down facilities have defaulted on their decommissioning funding obligation. Contingency plans are in place to ensure that decommissioning will be properly performed in the event of financial default of the licensee as discussed in the response to Question 10.14.

10.13 Who pays for decommissioning?

The particular licensee that holds the license for the facility pays for decommissioning. Subject to the public utilities commission that regulated the utility, the money for decommissioning is collected as part of the price of electricity; thus, the funds for decommissioning are ultimately paid by the ratepayer in the electric bill. As the electric utility industry deregulates, many states are choosing to require payment of decommissioning costs through the imposition of non-by-passable charge as part of a customer's electric bill.

10.14 What contingency plans are in place to ensure that decommissioning and long-term radioactive material storage will be properly performed in the event of financial default of the licensee? Who finances decommissioning if the licensee becomes bankrupt or insolvent?

The Atomic Energy Act contains provision for the Federal government to assume responsibility for decommissioning if public health and safety are jeopardized because of inability on the part of the licensee.

Bankruptcy does not necessarily mean that a power reactor licensee will liquidate. To date, the NRC's experience with bankrupt power reactor licensees has been that they file under Chapter 11 of the Bankruptcy Code for reorganization, not liquidation (for example, Public Service Company of New Hampshire, El Paso Electric Company, and Cajun Electric Cooperative). In these cases, bankrupt licensees have continued to provide adequate funds for safe operation and decommissioning, even as bondholders and stockholders suffered losses that were often severe. Because electric utilities typically provide an essential service in an exclusive franchise area, the NRC staff believes that, even in the unlikely case of a power reactor licensee liquidating, its service territory and obligations, including those for decommissioning, would revert to another entity without direct NRC intervention.

10.15 What will happen if deregulation becomes a reality? How will deregulation affect anticipated revenue and the capability to decommission?

The NRC has issued a final policy statement on its expectations and intended approach to nuclear power plant licensees as the electric utility industry moves from an environment of rate regulation toward greater competition. This policy statement was issued on August 19, 1997, and published in the *Federal Register* (62 FR 44071). The policy statement addresses NRC concerns about the adequacy of decommissioning funds. The statement indicates that the NRC believes that its current regulatory framework is generally sufficient to address the expected changes, but in order to remove any ambiguities in its regulations and address situations that may not be adequately covered, the Commission is considering revising its financial and decommissioning funding-assurance requirements.

Deregulation may force some licensees to separate their systems into functional areas, with their NRC-licensed nuclear plants potentially no longer being rate regulated. This would cause some licensees to cease being an "electric utility," as defined in NRC regulations. If this occurs, the NRC will require the licensees to meet the more stringent decommissioning funding-assurance requirements that apply to non-electric utilities. Electric utilities are permitted to accumulate funds for decommissioning over the remaining terms of their operating licenses. NRC regulations require most other licensees to provide funding assurance for the full estimated cost of decommissioning, either through full up-front funding or by some allowable guarantee or surety mechanism.

In addition, the policy statement emphasizes that the NRC retains the right to assess the timing of decommissioning trust fund deposits and withdrawals and the liquidity of decommissioning funds for licensees that no longer are subject to rate regulatory oversight.

The NRC issued a final rule on September 22, 1998, that modified its decommissioning funding-assurance requirements in anticipation of rate deregulation.

11 SOCIOECONOMIC ISSUES

11.1 What impact would each of the alternatives have on the economy of the surrounding area, including work-force requirements?

The biggest socioeconomic impact occurs before decommissioning starts, at the time the plant ceases operations, and the tax income created by the plant is substantially reduced. The surrounding communities may find their property tax base reduced in half or more, depending on the presence or absence of other industries in the area. Typically, additional public services are not required during decommissioning because the plant staff will be smaller than the operating staff.

11.2 How many people are needed onsite during the decommissioning process? Is this more than during operations?

After cessation of operations, the number of workers in the plant will be reduced. Plants that are currently being decommissioned using the DECON alternative have work forces in the range of 100 to 200 persons (for a single-unit plant). This is approximately one-third to one-tenth the number of persons who were employed at the plant during its operation.

These personnel are periodically supplemented with contract personnel during major decommissioning activities, such as the removal of large components like the steam generators and pressurizer. The onsite workforce could exceed 600 personnel during these peak periods of activity. If the plant were placed in SAFSTOR, the number of workers would be further reduced. Decommissioning plants that are located at the same site as operating facilities generally have a staff of 20 or fewer during SAFSTOR. Single-unit plants (not located next to operating units) require a larger staff and may have 20 to 70 employees during SAFSTOR. After the SAFSTOR period, the number of workers would increase to the range of 100 to 200 and would be further supplemented with contractor personnel for the final cleanup of the site.

11.3 What is being done to assist communities that relied heavily on the operating plant for their tax base, now that the plant has permanently ceased operations?

The NRC has no mechanism for providing economic assistance to communities that relied heavily on the operating plant for their tax base. This matter is generally left to local and State governments to address.

12 PUBLIC INVOLVEMENT

12.1 Public Meetings

12.1.1 What meetings are planned to keep the public informed?

Two public meetings are required during the decommissioning process. The first occurs before major decommissioning activities begin, when the post-shutdown decommissioning activities report is submitted. The second takes place when the licensee submit a license-termination plan, which describes how the site will be returned to a condition that makes radiological controls no longer necessary, must be submitted by the licensee. In both cases, the NRC will publish notifications of the public meetings in the *Federal Register* and in local media. The meetings will be held in the vicinity of the power plant to encourage local participation.

Although not required by the regulations, the NRC holds an initial public informational meeting shortly after the licensee submits the certification of permanent cessation of operations. At the meeting, the NRC presents its process for regulating decommissioning, the licensee presents its current plans for shutting down the facility and for decommissioning it (if any such plans have been made), and questions and comments from members of the public are addressed.

12.1.2 Where will the meetings be held?

Meetings are held in the vicinity of the facility. Often they are held at a local hotel, county courthouse, school, or library.

12.1.3 How will the public be notified about the meetings?

NRC issues a news release and publishes a notice of the date, time, and location of the public meeting in the *Federal Register* and in the local papers. The notice is also posted on the meetings section of the NRC's website (www.nrc.gov). See the response to Question 13.3 for additional information on accessing the NRC's website.

12.1.4 If I cannot attend a meeting, how do I find out what was said?

A written transcript of the meeting is prepared. A signup sheet is made available for members of the public to leave their addresses if they wish to receive a copy of the transcript. An individual who is unable to attend the meeting may contact the NRC project manager whose name, address, and telephone number are listed in the *Federal Register* notice and other published notices. Individuals may also call the NRC Office of Public Affairs, 1-301-415-8200, or the NRC project manager for the facility being decommissioned.

12.1.5 May I make comments at the meeting?

Yes. A portion of the meeting will be devoted to questions from the public. Additionally a sign-up sheet will be available at the start of the meeting for individuals requesting time for making comments or for reading prepared statements. Questions and short comments will also be accepted from the floor.

12.2   Comment Period

12.2.1 When does the comment period for the decommissioning process start?

There are two separate comment periods. The first occurs following the licensee's submittal of the post-shutdown decommissioning activities report (PSDAR) discussed earlier. A notice of the receipt of the PSDAR and the scheduling of a public meeting to be held in the vicinity of the licensee's facility is printed the *Federal Register*, posted in local places, and printed in a local newspaper. The PSDAR will be available for public comment at that time.

There is a second opportunity for public comment after the licensee submits a license termination plan which describes the remaining activities necessary to terminate the NRC license. Again, the NRC publishes a notice of the receipt of the license-termination plan in the *Federal Register*, and schedules a public meeting to be held in the vicinity of the licensee's facility. This notice is also published in the local vicinity of the site and in a local newspaper. Comments may be made in writing or orally at the public meeting.

12.2.2 How do I make comments on the decommissioning process?

Comments and questions may be submitted in writing to the NRC project manager for the facility. Comments and questions can also be addressed at the public meeting following receipt of the Post-Shutdown Decommissioning Activities Report (PSDAR) or the License Termination Plan. A written transcript containing these comments is prepared. All comments and questions received at the meeting will be responded to in a written memorandum that will be made available to the public. Additionally, a memorandum that documents whether or not the information provided in the PSDAR satisfies NRC requirements will also be prepared and made available to the public. A signup sheet will also be available at the public meeting for individuals to request copies of the memorandum, and a copy will be mailed to those who request one from the NRC project manager. An address and telephone number for the specific

NRC project manager will be published in the *Federal Register* along with the notice of the receipt of the PSDAR and the schedule of the public meeting.

12.2.3 Where should I send my comments?

NRC publications and notices identify the correct address and person to write to when submitting comments.

12.3 Hearings

12.3.1 Are hearings held on the decommissioning process?

Whenever there is an amendment to the facility's license, there is an opportunity for a hearing. For example, if the NRC finds the license-termination plan acceptable, it approves the plan by license amendment, which allows the opportunity for members of the public to request a hearing. The Post-shutdown Decommissioning Activities Report (PSDAR) is not a request for a license amendment, therefore, there is no opportunity at this stage in the process for a hearing on the contents of the PSDAR. The public meetings that follow the submittal of the PSDAR provide an additional opportunity for public participation.

12.3.2 Why is there no opportunity for a public hearing in the early stages of decommissioning?

The early stages of decommissioning, including the submittal of the Post-shutdown Decommissioning Activities Report, is not considered a major licensing action and does not, therefore, require a license amendment. If a license amendment is requested, there will be an opportunity for a public hearing.

12.3.3 What is the process that is used for submitting contentions for a public hearing?

The process for requesting or participating in a hearing on a license amendment is set forth in the NRC's regulations in Part 2, "Rules of Practice of Domestic Licensing Proceedings and Issuance of Orders" in the *Code of Federal Regulations*, Title 10, "Energy." Information on obtaining a copy of these regulations is given in the response to Question 13.6.

The regulations state that after publication of the proposed action (license amendment) in the *Federal Register*, any person whose interest may be affected by the proceeding may file a request for a hearing or a petition for leave to intervene if a hearing has already been requested. The *Federal Register* notice describes the time period in which this request or petition must be made. The person or group submitting the request or petition must explain in the petition (among other things) their interest in the proceeding and how that interest may be affected by the results of the proceeding. At a later time, they must also submit a list of contentions, which are specific statements of the issue of law or fact that are being raised. A brief explanation of the contention is required, which includes facts or expert opinion that supports the contention and sufficient information to show that a genuine dispute exists. Specific requirements for the information required in these documents and for the timing of the information submittals is specified in the NRC's regulations.

12.3.4 Does the public really have any power to make changes if there are problems?

Yes. There are several methods by which the public can make changes. If the licensee has requested an action requiring a license amendment, then the process for intervening in this action is discussed in the response to Question 12.3.3. If the action of concern does not involve a license amendment, any member of the public may raise potential health and safety issues in a petition to the NRC to take specific enforcement action against a licensed facility. This provision is contained in the NRC's regulations and is often referred to as a "2.206 petition" in reference to its location in the regulations (Chapter 2, Section 206 of Title 10, "Energy." In general, the petition is submitted in writing to the Executive Director for Operations and includes specific facts supporting the request for the NRC to take enforcement action. Unsupported assertions of safety problems or general opposition to nuclear power are not considered sufficient grounds for action. Further information on this approach is found on the NRC's website under Public Outreach (http://www.nrc.gov/OPA).

Workers at the nuclear power plant, or members of the public that have specific concerns of a safety-related nature, can bring safety concerns (allegations) directly to the NRC (although workers are encouraged to take their concerns directly to their employer initially). The NRC's toll-free safety hotline is 1-800-695-7403.

Members of the public may also voice concerns at the public meetings that are held in conjunction with the licensee's submittal of the Post-Shutdown Decommissioning Activities Report or License Termination Plan, or they may request information or identify concerns in writing. General correspondence should be addressed to the Office of Secretary, U.S. Nuclear Regulatory Commission, Washington, D.C. 20555-0001.

13 GETTING ADDITIONAL INFORMATION

13.1 Other than the public meeting, how can I get information about a nuclear power plant?

The NRC has many ways of keeping the public informed. These include printed materials, electronic access, and public meetings. NRC's website contains information of interest to the public (http://www.nrc.gov). This site also has a link to the Public Electronic Reading Room as described in the response to Question 13.2.

A comprehensive listing of information sources can be found in an NRC publication, NUREG/BR-0010, Rev. 2: *Citizen's Guide to U.S. Nuclear Regulatory Commission Information*.

The NRC maintains a Public Document Reading room in Washington, D.C. as discussed in the response to Question 13.4

Copies of documents can also be ordered from the Government Printing Office at the following address:

Superintendent of Documents
U.S. Government Printing Office
P.O. Box 37082
Washington, DC 20402-9328
Phone: 202-512-1800

Copies can also be read online or downloaded electronically from the NRC's website. The URL (address or Universal Resource Locator) for this is http://www.nrc.gov/NRC/NUREGS/BR0010/index.html.

13.2 What is the Public Electronic Reading Room, and how do I access it?

The Public Electronic Reading Room provides access to the NRC's new records-management system of publicly available information, the Agency wide Documents Access and Management System (ADAMS). Within this system you can access two libraries: the Publicly Available Records System, and that Public Legacy Library.

This system, which was implemented on October 12, 1999, marks a change in the previous practice where records were available only in paper or microfiche copies at either the main NRC Public Document Room in Washington, D.C. or at 86 local public document rooms at libraries near nuclear power plants and other regulated facilities throughout the United States. Access to the NRC Public Electronic Reading Room will now be possible from personal computers, including those located in most public libraries.

ADAMS is an electronic information system that allows access to NRC's publicly available documents via the Internet. It permits full text searching and the capability to view document images, download files, and print locally. It also provides a more timely release of information by the NRC and faster access to documents by the public than before.

ADAMS can be accessed via the Internet at the NRC's website using the following URL: http://www.nrc.gov/NRC/ADAMS/index.html. This site contains instructions for installing and running ADAMS as well as information on obtaining assistance during installation or use.

13.3 Does the NRC have a website? What kind of information can I obtain from it?

The NRC maintains a website at http://www.nrc.gov. The website includes

- general information about the NRC

- information on nuclear reactors (including the nuclear plant performance, plant information books, and daily reports, including plant status reports and daily events reports)

- information on radioactive waste disposal

- information on current rulemaking processes

- news releases and fact sheets on a variety of nuclear issues

- information on public involvement and public meetings, including sources for additional information

- copies of the post-shutdown decommissioning activities reports (PSDARs) for plants that have permanently ceased operations.

13.4 Does the NRC maintain a Public Document Room?

Yes. The Public Document Room contains a comprehensive collection of unrestricted documents related to NRC licensing proceedings and other significant decisions and actions. Among the wide variety of agency documents available to the public at the Public Document Room are NRC NUREG reports and manuals, transcripts and summaries of Commission meetings and NRC Staff and licensee meetings; existing and proposed regulations and rulemakings; licenses and amendments; and correspondence on technical, legal, and regulatory matters. Technical Reference Librarians are available to assist on-site users and those who call or write with information requests. The Public Document Room is located at 2120 L Street, N.W., Washington, D.C. It is open Monday through Friday from 7:45 a.m. to 4: 15 p.m. eastern time, except Federal holidays. Information and service to offsite users is available by calling (202) 634-3273 or 1-800-397-4209. The fax number is (202) 634-3343 and the email address is pdr@nrc.gov. The mailing address is U.S. Nuclear Regulatory Commission, Public Document Room, Washington, D.C. 20555.

13.5 What is the *Federal Register* and how can I get a copy of it?

The *Federal Register* is a daily publication announcing rules, policies, and other important actions of the Federal Government. Copies are available at many local libraries. Copies are also available at the NRC's Public Document Room in Washington, D.C. You may also search the *Federal Register* database on the Internet at http://www.access.gpo.gov/su_docs/aces/aces140.html.

13.6 How can I get a copy of the *Code of Federal Regulations*?

Title 10 of the *Code of Federal Regulations* (the section containing regulations enforced by the NRC) is available on the NRC's website under the reference library (http://www.nrc.gov/NRC/CFR/index.html). Most local libraries have personal computers available for free access to the NRC's website via the Internet. The entire collection of codes of Federal Regulations is often available at local libraries in the reference section. They are also available for purchase from the Government Printing Office (GPO) by credit card at 202-512-1800, Monday through Friday, 8 a.m. to 4 p.m. EST (fax 202-512-2233, 24 hours a day) or by check by writing to the Superintendent of Documents, Attn: New Orders, P.O. Box 371954, Pittsburgh, PA 15250-7954. For GPO Customer Service, call 202-512-1803.

13.7 How can I get answers to additional questions that were not addressed in this document?

The NRC has established a toll-free number, 1-800-368-5642, for general inquiries from the public concerning information on NRC activities. The NRC accepts calls from telephones equipped with a telecommunication device for the deaf (TDD) at a special main switchboard number, 301-415-5575. TDD numbers are also available in the NRC Library (301-415-5609).

The phone number for the Public Affairs Office located at NRC headquarters is 301-415-8200.

Regional public-affairs offices are also available to answer questions on specific nuclear facilities at

- Region I (Philadelphia) 610-337-5000

- Region II (Atlanta) 404-562-4400

- Region III (Chicago) 630-829-9500

- Region IV (Dallas) 817-860-8100

Questions may be addressed in writing to
    Office of Public Affairs
    U.S. Nuclear Regulatory Commission
    Washington, DC 20555-0001

Electronic inquiries can be sent to opa@nrc.gov.

14 BIBLIOGRAPHY

Federal Register Notices

"Clarification of decommissioning funding requirements," *Federal Register*, Vol. 60, p. 38235, July 26, 1995 (10 CFR Parts 30, 40, 50, 51, 70, and 72).

"Decommissioning of nuclear power reactors," *Federal Register*, Vol. 61, p. 39278, July 29, 1996 (10 CFR Parts 2, 50, and 51).

"Definitions – Decommissioning," *Federal Register*, Vol. 62, p. 39089, July 21, 1997 (10 CFR 30.4).

Electric utility industry; restructuring and economic deregulation; public health and safety effects; policy statement. *Federal Register* Vol. 62, p. 44071, August 19, 1997.

"Financial assurance requirements for decommissioning nuclear power reactors," *Federal Register*, Vol. 63, p. 50465, September 22, 1998 (10 CFR Parts 30 and 50).

"General requirements for decommissioning nuclear facilities," *Federal Register*, Vol. 53, p. 24018, June 27, 1988 (10 CFR Parts 30, 40, 50, 51, 70, and 72).

"Public notification and public participation," *Federal Register*, Vol. 62, p. 39089, July 21, 1997 (10 CFR 20.1405).

"Radiological criteria for license termination," *Federal Register*, Vol. 62, p. 39058, July 21, 1997 (10 CFR Parts 20, 30, 40, 50, 51, 70, and 72).

"Timeliness in decommissioning of materials facilities," *Federal Register*, Vol. 59, p. 36026, July 15, 1994 (10 CFR Parts 2, 30, 40, 70, and 72).

Generic Letters

GL 79-19, "Packaging of Low-Level Radioactive Waste for Transport and Burial."

GL 79-30, "Packaging, Transport and Burial of Low-Level Radioactive Waste."

Information Notices

IN 80-24, "Low-Level Radioactive Waste Burial Criteria."

IN 83-5, "Obtaining Approval for Disposing of Very-Low-Level Radioactive Waste – 10 CFR Section 20.302."

IN 85-92, "Surveys of Wastes Before Disposal From Nuclear Reactor Facilities."

IN 86-90, "Requests to Dispose of Very Low-Level Radioactive Waste Pursuant to 10 CFR 20.302."

IN 87-3, "Segregation of Hazardous and Low-Level Radioactive Wastes."

IN 88-8, "Chemical Reactions With Radioactive Waste Solidification Agents."

IN 88-16, "Identifying Waste Generators in Shipments of Low-Level Waste to Land Disposal Facilities."

IN 89-13, "Alternate Waste Management Procedures in Case of Denial of Access to Low-Level Waste Disposal Sites."

IN 89-27, "Limitations on the Use of Waste Forms and High Integrity Containers for the Disposal of Low-Level Radioactive Waste."

IN 90-9, "Extended Interim Storage of Low-Level Radioactive Waste by Fuel Cycle and Materials Licensees."

IN 90-31, "Update on Waste Form and High Integrity Container. Review Status, Identification of Problems With Cement Solidification."

IN 90-75, "Denial of Access to Current Low-Level Radioactive Waste Disposal Facilities."

IN 91-3, "Management of Wastes Contaminated With Radioactive Materials ('Red Bag' Waste and Ordinary Trash)."

IN 91-65, "Emergency Access to Low-Level Radioactive Waste Disposal Facilities."

IN 94-7, "Solubility Criteria for Liquid Effluent Releases to Sanitary Sewage Under the Revised 10 CFR 20."

IN 94-23, "Guidance to Hazardous, Radioactive and Mixed Waste Generators on the Elements of a Waste Minimization Program."

IN 96-47, "Recordkeeping, Decommissioning Notifications for Disposals of Radioactive Waste by Land Burial Authorized Under Former 10 CFR 20.304, 20.302 and Current 20.2002."

NUREG-Series Reports

NUREG-0586, "Final Generic Environmental Impact Statement on Decommissioning of Nuclear Facilities," August 1988.

NUREG-0945 (Draft), "Draft Environmental Impact Statement for 10 CFR Part 61," September 1981.

NUREG-0945, "Final Environmental Impact Statement for 10 CFR Part 61," November 1982.

NUREG-1101, "Onsite Disposal of Radioactive Waste," 1986.

NUREG-1307, "Report on Waste Burial Charges," Rev. 4, June 1994.

NUREG-1337, "Standard Review Plan for the Review of Financial Assurance Mechanisms for Decommissioning Under 10 CFR 30, 40, 70, and 72," Rev. 1, August 1989.

NUREG-1444, Supplement 1, "Site Decommissioning Management Plan," November 1995.

NUREG-1496 (Draft), "Generic Environmental Impact Statement in Support of Rulemaking on Radiological Criteria for Decommissioning of NRC-License Nuclear Facilities," Vols. 1 and 2, August 1994.

NUREG-1496, "Generic Environmental Impact Statement in Support of Rulemaking on Radiological Criteria for License Termination of NRC-Licensed Nuclear Facilities," Vol. 1, July 1997.

NUREG-1500, "Working Draft Regulatory Guide on Release Criteria for Decommissioning: NRC Staff's Draft for Comment," August 1994.

NUREG-1501, "Background as a Residual Radioactivity Criterion for Decommissioning," August 1994.

NUREG-1505, "A Nonparametric Statistical Methodology for the Design and Analysis of Final Status Decommissioning Surveys," August 1995.

NUREG-1506, "Measurement Methods for Radiological Surveys in Support of New Decommissioning Criteria," August 1995.

NUREG-1507, "Minimum Detectable Concentrations With Typical Radiation Survey Instruments for Various Contaminants and Field Conditions," August 1995.

NUREG-1520, "Standard Review Plan for the Review of a License Application for a Fuel Cycle Facility," 1995.

NUREG-1573, "Branch Technical Position on a Performance Assessment Methodology for Low-Level Radioactive Waste Disposal Facilities," April 1997.

NUREG-1577, Rev 1, "Standard Review Plan on Power Reactor Licensee Financial Qualifications and Decommissioning Funding Assurance (March 1999).

NUREG/BR-0111, "Transporting Spent Fuel: Protection Provided Against Severe Highway and Railroad Accidents," March 1987.

NUREG/BR-0241, "NMSS Handbook for Decommissioning Fuel Cycle and Materials Licensees," March 1997.

NUREG/CR-0130, "Technology, Safety, and Costs of Decommissioning a Reference Pressurized Water Reactor Power Station," June 1978 (Addendum 1, July 1979; Addendum 2, July 1983; Addendum 3, September 1984; Addendum 4, July 1988).

NUREG/CR-0672, "Technology, Safety and Costs of Decommissioning a Reference Boiling Water Reactor Power Station," June 1980 (Addendum 1, July 1983; Addendum 2, September 1984;

Addendum 3, July 1988; Addendum 4, December 1990).

NUREG/CR-5512, "Residual Radioactive Contamination From Decommissioning, Technical Basis for Translating Contamination Levels to Annual Total Effective Dose Equivalent," Vols. 1 and 2, October 1992.

NUREG/CR-5849 (Draft), "Manual for Conducting Radiological Surveys in Support of License Termination," June 1992.

NUREG/CR-5884, "Revised Analysis of Decommissioning for the Reference Pressurized Water Reactor Power Station," November 1995.

NUREG/CR-6174, "Revised Analyses of Decommissioning for the Reference Boiling Water Reactor Power Station," (Pacific Northwest National Laboratory), July 1996.

NUREG/CR-6232, "Assessing the Environmental Availability of Uranium in Soils and Sediments" (Pacific Northwest Laboratory), June 1994.

Regulatory Guides

DG 1067, "Decommissioning of Nuclear Power Reactors" (draft guide), June 1997.

DG 1071, "Standard Format and Content for Post-Shutdown Decommissioning Activities Report," December 1997.

RG 1.86, "Termination of Operating Licenses for Nuclear Reactors," June 1974.

Miscellaneous

Branch Technical Position, "Disposal or Onsite Storage of Thorium or Uranium Wastes From Past Operations," 46 FR 52601, October 1981.

"Branch Technical Position onsite Characterization for Decommissioning," 1994.

Commission Paper From EDO, Mr. J. Taylor, to Chairman Jackson and Commissioners Rogers and Dicus, "Resolution of Spent Fuel Storage Pool Action Issues," February 1, 1996.

FC 83-23, Policy and Guidance Directive, "Guidelines for Decontamination of Facilities and Equipment Prior to Release for Unrestricted Use or Termination of Byproduct, Source and Special Nuclear Material Licensees," November 1983.

"Guidelines for Decontamination of Facilities and Equipment Prior to Release for Unrestricted Use or Termination of Byproduct, Source and Special Nuclear Material Licensees," August 1987.

PG-8-08, Policy and Guidance Directive "Scenarios for Assessing Potential Doses Associated With Residual Radioactivity," May 1994.

SECY-94-145, "Increase of Tritium and Iron-55 Unrestricted Use Limits for Surface Contamination at Shoreham and Fort St. Vrain." May 1994.

NRC FORM 335
(2-89)
NRCM 1102,
3201, 3202

U.S. NUCLEAR REGULATORY COMMISSION

BIBLIOGRAPHIC DATA SHEET

(See instructions on the reverse)

1. REPORT NUMBER
(Assigned by NRC, Add Vol., Supp., Rev., and Addendum Numbers, if any.)

NUREG-1628
Final

2. TITLE AND SUBTITLE

Staff Responses to Frequently Asked Questions Concerning Decommissioning of Nuclear Power Plants

Final Report

3. DATE REPORT PUBLISHED

MONTH	YEAR
June	2000

4. FIN OR GRANT NUMBER

5. AUTHOR(S)

J.L. Minns, M.T. Masnik

6. TYPE OF REPORT

Technical

7. PERIOD COVERED *(Inclusive Dates)*

8. PERFORMING ORGANIZATION - NAME AND ADDRESS *(If NRC, provide Division, Office or Region, U.S. Nuclear Regulatory Commission, and mailing address; if contractor, provide name and mailing address.)*

Division of Licensing Project Management
Office of Nuclear Reactor Regulation
U.S. Nuclear Regulatory Commission
Washington, DC 20555-0001

9. SPONSORING ORGANIZATION - NAME AND ADDRESS *(If NRC, type "Same as above"; if contractor, provide NRC Division, Office or Region, U.S. Nuclear Regulatory Commission, and mailing address.)*

Same as above.

10. SUPPLEMENTARY NOTES

11. ABSTRACT *(200 words or less)*

This report through a question and answer format, provides U.S. Nuclear Regulatory Commission (NRC) staff responses to frequently asked questions on the decommissioning process for commercial nuclear power reactors for the 21st century. The questions were taken from a variety of sources over the past several years, including written inquiries to the NRC and questions asked at public meetings and during informal discussions with the NRC staff. In responding to the questions, the NRC staff attempted to provide the answers in a clear and non-technical form.

With the increase in the number of power reactors beginning the decommissioning process and significant changes that occurred in the regulations since 1996, the staff realized that there was a general lack of understanding of the decommissioning process and the risks associated with decommissioning. This document was developed in response to the staffs' concerns. The report contains a definition of decommissioning and a discussion of alternatives. It also provides a focus on decommission experiences in the United States and how the NRC regulates the decommissioning process. Questions related to spent fuel, low-level waste, and transportation related to decommissioning are answered. Questions related to license termination, the ultimate disposition of the facility, partial site release, socioeconomics and finances for completing decommissioning and hazards associated with decommissioning are also addressed. This document also provides reposes to questions related to public involvement in decommissioning as well as providing the public with sources for obtaining additional information on decommissioning.

12. KEY WORDS/DESCRIPTORS *(List words or phrases that will assist researchers in locating the report.)*

Decommissioning- nuclear power reactors
Decommissioning- process, sites, regulations, licensing, inspection programs
Spent Fuel pools- independent spent fuel storage installation (ISFSI)
Radioactive low level waste
Transportation
Licenses termination/Final Disposition of Facility
Hazards associated with Decommissioning
Finances
Socioeconomics
Public Involvement
Partial site release

13. AVAILABILITY STATEMENT

unlimited

14. SECURITY CLASSIFICATION

(This Page)

unclassified

(This Report)

unclassified

15. NUMBER OF PAGES

16. PRICE

Printed
on recycled
paper

Federal Recycling Program

NUREG-1628 — STAFF RESPONSES TO FREQUENTLY-ASKED QUESTIONS CONCERNING DECOMMISSIONING OF NUCLEAR POWER PLANTS — JUNE 2000

UNITED STATES
NUCLEAR REGULATORY COMMISSION
WASHINGTON, D.C. 20555-0001

1975 25 years 2000

SPECIAL STANDARD MAIL
POSTAGE AND FEES PAID
USNRC
PERMIT NO. G-67

www.ingramcontent.com/pod-product-compliance
Lightning Source LLC
Chambersburg PA
CBHW081835170526
45167CB00007B/2820